ABOUT THE AUTHOR

John C. Payne is a professional marine electrical engineer and surveyor. During his long maritime and offshore oil career, he has served on a variety of commercial ships, from general cargo, reefers to oil tankers and dive support vessels, both as an Electrical Officer and Marine Engineer. He has served as a Diving Technician and is a Merchant Marine war veteran. In the offshore oil drilling industry, he is still actively involved with the world's most technologically advanced deepwater drilling rigs as a Construction and Commissioning Manager. John has been employed by one of the leading maritime consultancy companies as a surveyor on a wide variety of vessels, and acted as electrical expert witness on several large marine litigation cases.

As a professionally qualified technical author and writer, John has been responsible for defining and writing technical and operations manuals on a diverse range of air force, marine, and naval projects including naval research vessels and submarine sonar systems. He is a licensed electrical contractor and a Certified Maritime Safety Auditor under the ISM Code. Previously he has been a Fellow of the Institute of Diagnostic Engineers and the Institute of Scientific and Technical Communicators.

John is the author of *Marine Electrical and Electronics Bible*, *The Motorboat Electrical and Electronics Manual*, the Understanding Boat series, *The Fisherman's Electrical Manual*, and *Piracy Today*. As his passion for cruising is local cuisine, he also authored *The Great Cruising Cookbook*, or *Cruisine©*, as he likes to call it. He has been widely published around the world in various boating and yachting magazines and regularly lectures on the subject.

His sailing career started in high-performance racing dinghies. He has restored, cruised, and lived aboard a classic 37-foot Herreshoff ketch and a 34-foot wooden sloop cruising Europe, the UK, the Pacific, Australia, and the Mediterranean. After several years living aboard his 115-year-old Dutch barge navigating European rivers and canals, he is now back sailing and cruising aboard a 36-foot ketch. He is a member of the American Boat and Yacht Council (ABYC), the Cruising Association, and the Westerly Owners Association. His website is www.fishingandboats.com; follow him on Twitter/X @JohnCPayne55579.

ALSO BY JOHN C. PAYNE

Marine Electrical and Electronics Bible: 4th Edition

The Motorboat Electrical and Electronics Manual

The Fisherman's Electrical Manual: A Complete Guide to Electrical Systems for Bass

Boats and Other Trailerable Sport-fishing Boats

Understanding Boat Batteries and Battery Charging

Understanding Boat Communications

Understanding Boat Corrosion, Lightning Protection, and Interference

Understanding Marine Diesels: 2nd Edition

Understanding Boat Electronics

Understanding Boat Wiring: 2nd Edition

Understanding Boat Plumbing and Water Systems

The Great Cruising Cookbook: An International Galley Guide

Piracy Today: Fighting Villainy on the High Sea

UNDERSTANDING MARINE DIESELS

Second Edition

A ◂ SHERIDAN HOUSE ▸
Guide to Boat Maintenance

JOHN C. PAYNE

Essex, Connecticut

An imprint of Globe Pequot, the trade division of
The Rowman & Littlefield Publishing Group, Inc.
4501 Forbes Blvd., Ste. 200
Lanham, MD 20706
www.rowman.com

Distributed by NATIONAL BOOK NETWORK

First Sheridan House edition published 2005
Second edition published 2024

British Library Cataloguing in Publication Information available

Library of Congress Cataloging-in-Publication Data
Names: Payne, John C., 1954– author.
Title: Understanding marine diesels / John C. Payne.
Other titles: Understanding boat diesel engines
Description: Second edition. | Essex, Connecticut : Lyons Press, 2024. | Includes index. | Summary: "Concise, compact, and fully illustrated for easy reference, this fully revised guide offers a comprehensive coverage of marine diesel engine parts and what they do, checklists for regular engine care and maintenance as well as troubleshooting, and an overview of electrical systems"— Provided by publisher.
Identifiers: LCCN 2023042034 (print) | LCCN 2023042035 (ebook) | ISBN 9781574093599 (paper ; alk. paper) | ISBN 9781574093605 (electronic)
Subjects: LCSH: Marine diesel motors—Maintenance and repair—Handbooks, manuals, etc. | Boats and boating—Electric equipment—Handbooks, manuals, etc.
Classification: LCC VM771 .P38 2024 (print) | LCC VM771 (ebook) | DDC 623.87/236—dc23/eng/20231117
LC record available at https://lccn.loc.gov/2023042034
LC ebook record available at https://lccn.loc.gov/2023042035

♾™ The paper used in this publication meets the minimum requirements of American National Standard for Information Sciences—Permanence of Paper for Printed Library Materials, ANSI/NISO Z39.48-1992.

CONTENTS

ABBREVIATIONS AND ACRONYMS

AC alternating current

API American Petroleum Institute

ASTM American Society of Testing and Materials

BDC bottom dead center

CAN controller area network

CE conformité européenne

PU central processor units

DC direct current

DI direct injection

DZR dezincification-resistant brass

ECM electronic control module

ECM engine control module

ECU engine control unit

EI Energy Industry

EPDM ethylene propylene diene monomer

FW freshwater

gph gallons per hour

HEUI hydraulically actuated, electronic controlled, unit injector

IAT inorganic acid technology

IDI indirect injection

LED light emitting diode

lph liters per hour

mpg miles per gallon

mV millivolts

PDT pressure diffusion technology

ppm parts per million

PSU power supply unit

PTFE polytetrafluoroethylene

PWM pulse width modulated/modulation

RCD European Recreational Craft Directive

RFI radio frequency interference

RTD resistance temperature detector

SO_2 sulfur dioxide

SO_3 sulfur trioxide

SOLAS International Convention for Safety of Life at Sea

SW seawater

TBN total base number

TDC top dead center

ULSD ultra-low sulfur diesel

VG viscosity grade

Ω ohm

INTRODUCTION

The marine diesel engine is the main propulsion source for most cruising and sailing yachts, trawler yachts, and many powerboats. In the marine diesel space, the most common engines are Volvo Penta, Yanmar, Nanni, Detroit, Westerbeke, Beta Marine, Styer, Solé Diesel, Caterpillar, Perkins, Mercedes, Cummins, and Vetus Diesel. Most smaller yacht diesel engines are marinized engines, often from proven industrial and agricultural diesel engines that include Kubota, Mitsubishi, Toyota, and Shibaura. Marine diesel engines are now required to comply with stringent emissions standards, and these include the US EPA Domestic Marine Tier 2 and Tier 3 standards and the European Recreational Craft Directive (RCD).

1

Basic Diesel Theory

The diesel was invented back in 1892. It is named for pioneer German engineer Rudolf Diesel, who demonstrated the first such engine using peanut oil at the World Fair in Paris in 1900. The diesel engine works on the principle of compression-ignition, where air is compressed to a point where fuel combustion will occur spontaneously. This book covers four-stroke compression-ignition engines.

THE FOUR DIESEL ENGINE CYCLES

A diesel engine has four distinct phases to each cycle. The four-stroke cycle comprises the following:

1. **Air Intake (Induction) Stroke.** This is when the piston moves down and the air is drawn into the cylinder. At top dead center (TDC) the inlet valves open and the air required for fuel combustion is drawn in through the air inlet manifold, air filter, and turbocharger as the piston moves downward. At the bottom of the stroke, bottom dead center (BDC), the inlet valves close, sealing the cylinder full of fresh air.
2. **Compression Stroke.** In the compression stroke the piston moves upward to compress the air and raises the air temperature within the engine cylinder, typically to around 1,025°F (550°C). At just before TDC, fuel injection takes place and, after an interval, ceases.
3. **Power Stroke.** The fuel is injected into the heated and compressed air within the cylinder as a stream of small droplets. It ignites spontaneously when the droplets mix with the heated air. Once ignited, combustion occurs, and increased pressure is then generated in the cylinder. This explosively forces or drives the piston downward to BDC, which produces the power to turn the crankshaft and, subsequently, the shaft and propeller.

4. **Exhaust Stroke.** At BDC the exhaust valves open to expel exhaust gases. The piston travels back up, forcing the exhaust gases out through the valves into the exhaust system. At the end of the upward stroke, the valves close at TDC. This process is sometimes called gas exchange. The exhaust valves are controlled by push rods, a camshaft, and the crankshaft, and this takes place every second engine revolution. The camshaft rotates at half the speed of the crankshaft.

COMBUSTION EFFICIENCY

Combustion efficiency is a measure of compression ratio, and is highly dependent on the proper control of both ignition and fuel combustion. Factors controlling combustion are air quantity, fuel-air mixture, and compression temperature and pressure. The important factor is the delay period between the fuel injection and the ignition. Both engine design and fuel quality are crucial to this. It affects engine performance, cold-start characteristics, warm-up times, engine power output, engine noise, and the level of exhaust emissions. Short ignition delays do not generally cause problems; however, long ignition delays will allow fuel to accumulate in the cylinder prior to ignition. When this occurs, the cylinder pressure will rise rapidly, with incomplete and inefficient combustion. Ignition delay periods are mainly determined by fuel quality. Injection delay is the time required for the injection pump to build up pressure exceeding the opening pressure of the injector. This is dependent on fuel quality, compression temperature, compression pressure, and fuel droplet size. The ignition delay must be as short as possible. When ignition starts, combustion occurs quickly and the pressure increases rapidly. When delay is excessive, pressure rise is also fast, causing engine "knocking." During the final part of the combustion process, the final fuel is burned off. When the temperatures and pressure are high, the fuel droplets ignite immediately. Good combustion efficiency ceases at this point.

ENGINE EFFICIENCY

Efficiency and losses in the combustion process relate to fuel energy, incomplete combustion, and air/fuel mixing less than 100% so that unburned fuel exits in the exhaust gases. Losses are typically around 65% and are higher at low-load than full-load operation. Sources of losses include pressure leakage through piston rings, friction, heat loss through combustion chamber walls, incomplete expansion, thermal loss via the exhaust gases, and incomplete combustion. The air temperature, air humidity, and air pressure all affect the power ratings of an engine. In many small engine spaces, engine power is often reduced due to insufficient air supplies. The higher the engine space temperature rises, the greater the efficiency loss, which can be around 10% to 15%. Efficiency can be improved by installing an air inlet ducted to the engine air inlet, which provides cool external air. This is in addition to the normal machinery space ventilation inlets and outlets or exhaust fans.

DIESEL ENGINE SUBSYSTEMS

The principal subsystems of marine diesel engines are summarized below. Always consider each as a separate system, as each requires different maintenance and troubleshooting tasks. Eventually the collective sum of all these subsystems provides the motive power for your vessel, along with power generation.

Main Engine Components

The engine comprises the block, the cylinder head, the head gasket, and the top cover. The block supports the crankshaft, camshaft, cam followers, cylinders, and pistons. Inlet manifolds and exhaust manifolds are attached. The top end comprises valve lifters, push rods, rocker arms, tappets, timing gears, valve springs, and valves. The piston assembly comprises components that include the piston pin, upper and lower compression rings, and oil rings. The bottom end of the assembly comprises the connecting rods, bolts and nuts, main or rod bearings, and rod cap.

The Diesel Fuel System

The newer diesel engine technologies have electronic control with sensors and timing, along with design innovations that include improved fuel pumps, nozzles, and unit injectors. The diesel fuel injection system causes the most problems, through either a lack of maintenance or the use of low-quality fuel. This can result in poor fuel economy, excessive exhaust smoke, excessive carbon accumulations within the combustion chamber, and shortened engine life. Low-quality fuels can cause starting problems and engine knocking. Engines have high injection pressures, and fuel pumps and injectors have very close tolerances for optimum fuel atomization and penetration. The diesel fuel must have lubrication capabilities to prevent excess wear and resultant damage.

The Cooling System

The cooling system encompasses the seawater system, the closed freshwater system, and the heat exchanger. This includes strainers, water pumps, and thermostats. The correct temperature-regulating function of the thermostat is important to efficient operation. A diesel engine running cold is almost as bad as an engine running too hot. The cooling system needs regular monitoring and maintenance.

The Lubrication System

The engine lubrication system is crucial to the efficiency, cooling, and longevity of the engine. The oil pump supplies the oil from the sump through the engine block passages and galleries to all the engine's working surfaces and parts.

The Air System

The air system is crucial to good engine performance, and, unlike their land-based cousins, clogged air cleaners and filters are not the prime cause of problems afloat. Instead, issues arise when there is an inadequate fresh air supply.

The Exhaust System

This is the cause of major issues when improperly installed, particular wet exhausts. Exhaust smoke color can be a very reliable indicator of engine performance.

The Starting System

Probably the most common marine diesel engine system issue is the starting system. There are a number of factors to consider, including the starter motor and solenoid performance.

The Instrumentation System

This is a system of meters and sensors for monitoring basic parameters, from engine speed to oil pressure and water temperature, with associated alarms for out-of-tolerance operating conditions.

The Charging System

Most marine diesel engines rely on the engine charging system alternator to recharge their own dedicated start battery and house power battery.

The Engine Mounts

Engine mounts absorb and dampen engine vibrations. Some shock-absorbing mounts, such as the VETUS hydro-dampener units, significantly reduce noise and vibration. Often forgotten, they do fatigue and need inspection.

2

The Fuel System

Engine combustion efficiency is significantly influenced by the design of the fuel system. Fuel is drawn from the fuel storage tank, through a pre-filter or separator and the engine fuel filter, to the fuel pump. The pressurized fuel is then fed to the injectors. Although fuel is injected as a liquid, efficiency depends on the correct vaporization. This requires thorough and rapid mixing of both the hot compressed air and the fuel vapor. Injection pressures will range from 2,500psi to 7,000psi (172bar to 482bar).

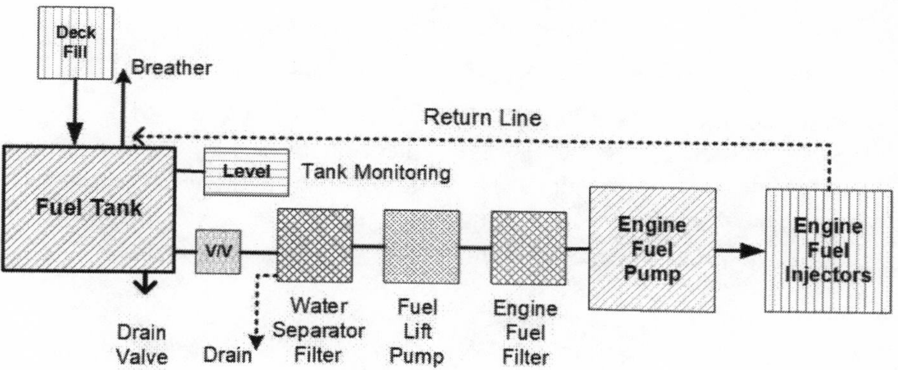

Figure 2-1. Marine Diesel Fuel System

MARINE DIESEL FUEL SYSTEM COMPONENTS

The typical diesel fuel system comprises several main functional components:

1. **Fuel Oil Tanks.** This is the main storage tank or tanks, with diesel fuel being loaded or bunkered through the deck filler. The deck filler cap must always be tight, seal well, and be watertight. The fuel tank should have a vent or breather to prevent pressure or partial vacuum formation. This must be located above the water line and prevent water ingress. The tank should be angled so that water and sediment can collect and be drained off at the low point through a drain cock. This should be done regularly. Fuel tanks should be kept as full as possible at all times to reduce condensation. Being full reduces the amount of fuel movement or slopping that causes aeration or stirring up of debris particles trapped within a tank.

2. **Header or Day Tank.** Some boats have a header tank where fuel is transferred from the main storage tank. This tank is often mounted higher than the engine to allow a positive gravity feed to the engine.

3. **Fuel Pre-filter/Separator.** This is installed in the fuel suction line before the diesel fuel enters the engine. Typical of these are the Parker Racor and Baldwin-Dahl type units. These trap oil and water, and filter particles and debris before it reaches the engine filters. These are easily drained through a plug and monitored through a transparent bowl.

ABOUT THE ENGINE FUEL SYSTEM

1. **Fuel Lift Pump.** The main role is to ensure adequate fuel supplies for combustion. This pump is used to bleed the engine with the integral manual operating lever where installed.

2. **Engine Fuel Filter.** This is a disposable fine filter mounted on the engine. It is essential that the fuel element is kept clean, as this is the prime component in maintaining fuel oil quality. Always fill the filter with fuel if possible and smear some fuel oil on the rubber sealing ring when renewing the disposable element.

3. **Fuel Injection Pump.** This pump pressurizes the fuel to the injectors. It meters the correct fuel quantity at the precise time it is required at the top of the compression stroke. The pump is a precision piece of equipment, and both dirt and water will seriously damage it.

4. **Injectors.** These inject atomized diesel fuel into the cylinder. The injector is a precision piece of equipment. The injector will spray the fine fuel mist into the cylinder in a designated spray pattern to properly distribute fuel for optimum ignition and combustion efficiency.

5. **Governor.** This controls the engine speed during load variations. The governor regulates the injection pump fuel output to the injectors. These can be mechanical or electronic.

HOW DIESEL FUEL INJECTION WORKS

The quantity of diesel fuel delivered at each injection is controlled by the injection pressure, the injector nozzle orifice area, and the time at which the nozzle valve lifts. To increase the vaporization rate, the fuel must be injected into the cylinder as an atomized stream of fuel droplets. The fuel injector must meter the correct quantity of fuel to match the engine power requirements, and the injection period must be determined precisely. The droplet size of the fuel is critical to achieving good combustion. Large fuel droplets will require a longer period to vaporize, delaying combustion. Small fuel droplets move relatively slowly, reducing the oxygen mixing times. Both conditions will lead to incomplete combustion, with reduced efficiency, increased noise, and increased emissions.

FUEL INJECTOR CONSTRUCTION

The injector consists of several components that include the tappet, plunger, barrel, body, nozzle assembly (which includes the spring, check, and tip), and cartridge valve (which consists of a solenoid, armature, poppet valve, and poppet spring). In mechanical systems, push rods and cam lobes activate rocker arms and the injector plunger and barrel. In electronic systems, the engine control module (ECM) energizes the solenoid. This magnetically attracts the armature and lifts the poppet valve and allows fuel pressure to build up; the check lifts and is then injected via the nozzle assembly into the cylinder. At the programmed end of injection, the solenoid valve de-energizes and fuel flow ceases.

ABOUT COMBUSTION EFFICIENCY

Combustion is significantly influenced by the design of the fuel injection system. Fuel is drawn from the fuel tank, through pre-filters or separators and the engine fuel filter, to the fuel pump. The pressurized fuel is then supplied to the injectors. Fuel is injected at high pressure. Although fuel is injected as a liquid, efficiency depends on optimal vaporization, and this requires thorough and rapid mixing of both the hot air and the fuel vapor. There are several distinct fuel injection methods: the constant-pressure common rail system, individual pump system, multiple plunger system, in-line (jerk) system, accumulator system, pressure-time injection system, and distributor pump system.

CONSTANT-PRESSURE INJECTION SYSTEM

In the constant-pressure common rail system, the fuel is maintained at a constant pressure in the manifold. The manifold is connected to cam-actuated nozzles or a distributor, the timing valve, and pressure-activated injector nozzles. The pressure is maintained by compressing the diesel fuel using a pump and supplying fuel after each injection. Fuel is supplied from an accumulator and pressure regulating valve, which may be a governor or manually controlled. The quantity of fuel delivered at each injection is controlled by injection pressure, the injector nozzle orifice area, and the time at which the nozzle valve lifts.

ABOUT THE ACCUMULATOR INJECTION SYSTEM

The accumulator-type injection system uses both upper and lower plungers in a common bore. The lower plunger is driven by an eccentric cam; the upper plunger is spring-loaded. As the bottom plunger is forced up, the fuel between the plungers is pressurized. This is dependent on the spring force applied to the top spring. Fuel will continue to pressurize until a delivery groove in the lower plunger indexes with the outlet passage. The pressurized fuel is then injected and continues until the upper spring forces the plunger downward and closes the outlet passage.

ABOUT THE JERK PUMP INJECTION SYSTEM

The jerk pump injection system is the most common system in use for fuel pressurization, metering, and timing. A camshaft is used to activate the plungers and control injection. The jerk pump system is the basis for distributor pumps and unit injectors. Engine makers such as Caterpillar use hydraulically actuated, electronically controlled, unit injectors (HEUIs), which have fuel injection pressures in the range 18,000psi to 24,000psi (1200bar to 1650bar).

ELECTRONIC FUEL INJECTION

Each injector has a solenoid valve that controls the quantity of fuel to the injector. The gear-driven axial pump raised the fuel oil pressure for proper injection operation. The ECM transmits a signal to an injection pressure control valve, or dump valve, and a signal to each injector solenoid valve to inject the fuel. The control valve controls the injection pump outlet pressure by dumping oil back. Yanmar common rail technology utilizes a digitally controlled, high-pressure fuel injection and sensor system that reduces emissions. The system has multiple engine sensors that monitor operating parameters back to the engine control unit (ECU). This includes throttle position, various temperatures, oil, water, common rail and other pressures, and intake air pressures, as well as crank and cam positions. The ECU uses this data to regulate fuel from the pump into the high-pressure fuel rail and digitally activated injectors in order to precisely control fuel injection.

ABOUT INDIRECT INJECTION (IDI)

In the IDI engine, the air is drawn in through a shaped inlet port or prechamber connected to the cylinder top housing the fuel injector. This causes the air to swirl during compression, and combustion starts here. This slows the combustion rate and produces less noise. The IDI engine has greater fuel economy, and turbocharging can enhance this. An increased useful speed range is obtained as larger valves are used, and noise levels are reduced.

ABOUT DIRECT INJECTION (DI)

The diesel fuel is directly injected into the cylinder. The piston incorporates a toroidal depression where combustion occurs. These engine types are the most efficient, and can deliver high power over short time periods. They have a relatively low acceleration, as both the inlet and exhaust valves are restricted in diameter to allow for a centrally located injector, which inhibits the aspiration process. This results in better economy and lower exhaust emissions.

ABOUT THE FUEL PUMPS

The fuel injection pump is often labeled the heart of the diesel engine. High-pressure fuel pumps are capable of producing anywhere from 2,000psi to 35,000psi (140bar to 2,400bar) or more, depending on the engine and system design. The different types include the continuous pump, the distributor pump, and the individual pump. Pumps might be an electric lift pump or a cam-operated lift pump. Common problems are leaks from seals and gasket failure. Poor fuel quality or filtering can coat internal components with gum, and running fresh fuel through the system is important.

THE FUEL SYSTEM CYCLE EXPLAINED

There are five functions within a diesel engine fuel system:

1. **Fuel Metering.** Fuel metering must be accurate so that, at the same fuel control setting, the same quantity of fuel is delivered to each cylinder for each power stroke. This enables consistent speed and power output, ensuring even power distribution between each cylinder.

2. **Fuel Injection Control.** The rate of fuel injection is critical, as this determines the combustion rate. At the start, it is important that excess fuel does not accumulate within the cylinder during the initial time delay. Injection should be set so that combustion pressure is not excessive and fuel injection happens at a rate to obtain complete combustion.

3. **Timing.** The system must synchronize injection along with metering the correct quantity of fuel. This is to ensure efficient combustion and energy creation. If fuel is injected too early, ignition delay occurs. Excessive delay results in noisy and rough engine operation. Fuel is often wasted due to the wetting effect on the cylinder walls and the head of the piston. This can result in diluting the lubricating oil with diesel fuel.

4. **Fuel Atomization.** Atomization is the process of converting the pressurized fuel into small particles or droplets to form a mist. The mist or spray should be optimized for the combustion chamber. Different chamber designs have different atomization patterns. Each spray particle should be surrounded by oxygen particles to ensure proper and even combustion. When the high-pressure fuel passes into the injector and through the nozzle, it is forced through the injector nozzle holes at the spray tip. Friction then assists in breaking the fuel stream into particles, or micro-droplets.

5. **Pressure Creation.** The fuel injection process is required to increase the fuel pressure to a level greater than the compression pressure. This is required to ensure the correct level of fuel dispersion and an even combination of air and fuel for efficient combustion. If the fuel particles are too small, they will not penetrate the chamber for even and complete combustion.

ABOUT DIESEL FUEL QUALITY

The quality of diesel fuel is critical to good engine combustion. The US standard for diesel is the American Society of Testing and Materials (ASTM) D 975-21. These are classified as No.1-D and No.2-D. No.2-D is used in boats and vehicles and, unfortunately, often has a reputation as being of variable quality and purity. This is attributable to substandard storage and handling after it leaves the refinery. No.1-D is blended for cold temperature performance. The UK and Europe adhere to the EN590 standard. The Environmental Protection Agency (EPA) has mandated cleaner-burning fuels for reduced emissions, which has led to ultra-low sulfur diesel (ULSD), with a maximum of 15ppm sulfur content. The sulfur content is crucial in controlling exhaust emissions. In the combustion process, the sulfur compounds alter and mix with water to form acidic by-products, notably sulfur dioxide (SO_2) and sulfur trioxide (SO_3), which enter the exhaust gas, causing higher emission pollution. High sulfur levels are corrosive, and will cause high engine wear rates and degrade engine oil additives. The quality of diesel fuel impacts engine starting characteristics, the wear rate on fuel injection equipment, and the wear on pistons, rings, valves, and cylinder liners.

ABOUT CETANE NUMBERS

Another prime factor is the Cetane number, which is the measure of oil volatility. The higher the Cetane rating, the easier the engine starts and the more combustion efficiency improves. No.2-D diesel has a Cetane rating in the range of 40 to 50. The Cetane number affects the ignition delay period. This is crucial to engine starting, often causing white smoke after startup, diesel knock at idle speeds, and lower overall engine performance. Choose the highest standard fuel you can—while initially more expensive, it results in lower maintenance and running costs.

ABOUT DIESEL FUEL CONTAMINANTS

The American Petroleum Institute (API) specifications allow for acceptable levels of impurities such as sulfur, wax, and other contaminants, including water, dirt, and ash. Fuel contaminates are classified as either precipitates or particulates. The precipitates are commonly noncombustible materials that form when fuel oxidizes. As they are heavier than fuel, they normally fall to the bottom of fuel tanks. Particulates are often known as asphaltenes, black tar-like substances that can plug fuel filters. Particulates tend to be suspended within the fuel and take time to settle out.

ABOUT THE INJECTION DELAY PERIOD

An important factor in combustion efficiency is the delay period between the fuel injection and the ignition. This is dependent on the fuel quality, compression temperature, compression pressure, and fuel droplet sizes. Injection delay is the time required for the injection pump to build up pressure exceeding the opening pressure of the injector, and must be as short as possible. Both engine design and fuel quality are crucial to this, and it affects engine performance, cold start characteristics, warm-up times, engine power output, engine noise, and level of exhaust emissions. Short ignition delays do not generally cause problems. Long delays will allow fuel to accumulate in the cylinder prior to ignition. When this occurs, the cylinder pressure will rise rapidly, with incomplete and inefficient combustion. When delay is excessive, the pressure rise is fast, which can cause "knocking." During the final part of the combustion process, the final fuel is burned off. When temperatures and pressure are so high, the fuel droplets ignite at once. Good combustion efficiency ceases at this point.

DIESEL FUEL LUBRICITY EXPLAINED

A fuel system has moving parts in the fuel injectors and fuel pumps. These surfaces require lubrication to reduce friction and wear, and this is a function of diesel fuel. Diesel fuels are blended to ensure engines have adequate lubricity. Many use additives to improve fuel characteristics, and lubricity is one of them.

DIESEL FUEL VISCOSITY EXPLAINED

The most important factor, assuming that the fuel is clean, is the viscosity. Fuel viscosity affects the flow of the fuel. Very high fuel viscosities impose severe strains on the fuel system, in particular the fuel pump and injectors. Low viscosities will cause leakage past the fuel pump, as well as cause unnecessary wear of vital components, as fuel acts as a lubricant. Viscosity will impact fuel combustion after injection, as it can affect atomization or fuel droplet size, which affects the spray pattern.

DIESEL FUEL CLOUD AND POUR POINTS

The cloud point is the temperature where wax crystals within the paraffin base of the fuel start to settle out, which results in clogged fuel filters. This is encountered in cold temperature environments. Some boats, like some vehicles, have fuel heaters with thermostatic controls to counter the problem. The pour point of the fuel is the lowest temperature at which fuel can be pumped and circulate through the system. The pour point is 5°F (2.8°C) greater than the point at which diesel fuel will not flow or starts to solidify. The cloud point of the fuel is typically around 9°F to 14°F (5°C to 7.7°C) above the pour point.

ABOUT WATER IN DIESEL FUEL

Water in diesel fuel is the great enemy for all marine diesel engines. Water is the worst and most common dangerous contaminant in fuel, destroying the fuel's lubrication qualities and damaging fuel pumps and injectors. During transportation and storage, fuel is in contact with air and water vapor, and can undergo a change in quality as a result. Sometimes sludge is formed, which may block the fuel filters, or gums are created that can damage the fuel injection equipment or leave deposits within the engine. Moisture also causes corrosion and can result in injector seizure. There are several measures to reduce condensation in diesel fuel:

1. **Fuel Tanks.** Fuel tanks should always be topped up to reduce condensation; where possible, tanks with a drain valve should have water drained off regularly. If you don't have a drain valve at the lowest point of your fuel rank, consider installing one.
2. **Post Filling.** After filling diesel fuel tanks, the pre-filters and separators should be monitored for excess water and drained. The single greatest investment that you can make with the fuel system is to install a filter/separator unit, also called sedimenters and agglomerators. The main manufacturers include Parker Racor, Baldwin-Dahl, Delphi CAV, Keenan, and Vetus.

MICROBE GROWTH EXPLAINED

Moisture and water within the system can encourage microbial growth within the fuel, including algae, fungi, and bacteria (both aerobic and anaerobic), which multiply rapidly and plug the filters. Microbial growth is highly dependent on temperature, thriving in the 50°F to 115°F (10°C to 45°C) range. Once the system is "infected," considerable flushing is required if the infection is ever to be eliminated. The solution is to add chemical biocides to kill the contaminant or maintain quality. One such option on the market are devices called magnetic biocides, such as the De-Bug. The theory is that single-celled organisms have electrical potential, usually positive on the cell wall with a negative interior. If this is disrupted, the cell will die or rupture. This can be initiated by passing fuel over a strong permanent magnet or series of magnets. In the De-Bug unit, the cells are exposed to many magnetic field changes within a short distance of travel. The fluctuations destroy the cells by disruption of cell ions and pH balances.

DAHL FILTER OPERATION

1. Fuel from the diesel tank enters the filter inlet port and is directed down through the center tube. The de-pressurizer cone then spreads the fuel.
2. As fuel is discharged from the de-pressurizer cone, 80% of the contaminants are separated from the fuel. The fuel rises upward, and most of the solid contaminants and water settle into the bowl's quiet zone. The system includes a reverse-flow valve to hold prime in the fuel system and prevent fuel from flowing back to the tank during shutdown. There is a removable primer plug at the top, for use when complete priming is required.
3. As the fuel rises upward, remaining small water droplets collect on the cone, baffle, and bowl surfaces. The size and weight of the water droplets gradually increase, causing downward flow into the sump.
4. Fuel is then filtered completely by the 2-micron paper element. The clean fuel continues up through the outlet port to the pump and injection system. The transparent bowl holds up to 24 fluid ounces of water capacity to reduce draining intervals; this is done via the drain cock.

PARKER RACOR FILTER OPERATION

The units have optional heaters, electrical water continuity probe alarms, and a vacuum gauge to monitor pressure drop across filter elements. I have used these on my past boats.

1. In the separation stage, a turbine separates large solids and free water using centrifugal force.
2. In the coalescing stage, the smaller water droplets and solids coalesce on a conical baffle and drop into the collection bowl.
3. The filtration stage uses a fine-micron AquaBloc water-repelling paper element.

ABOUT VETUS FILTERS

These are spin-on-type filters based on a patented full-fuel flow system that separates water and particles prior to fuel flow through the filter element. They have a CE (ISO10088)- and American Boat and Yacht Council (ABYC)-compliant clear bowl. For offshore sailing, a dual filter system is recommended, as rough weather motions generally result in tank agitation and increased input of particles. Dual systems have a vacuum gauge to indicate filter clogging.

ABOUT THE KEENAN FILTERBOSS UNIT

These are either a single or dual filter and a polishing system. The single-element MK60SP has a remote warning panel to alert system problems and an integrated fuel pump for bleeding, priming, and backup of the main engine fuel pump. It uses Racor filters, which are easy to obtain. The unit has a vacuum gauge that's color coded to determine if operation is normal. The integral pump allows a return line to the tank to polish stored fuel. The system has optional smartphone connectivity.

DELPHI CAV UNIT

The fuel filter housing has a 5 micron filter, and water is collected in the glass bowl.

ABOUT THE DIESEL DIPPER

This patented fuel treatment system from Marine 16™ reimagines commercial ship purifier systems for yacht applications. The system suctions fuel a half to 1in (10-25mm) above the fuel tank bottom with a small 12V pump. At this level, any water and sediment that contributes to microbial growth is removed. The extracted fuel passes through a swirl chamber and filter to separate out contaminants. It is then collected into a small storage or settling tank that has a drain valve for removal of the collected water and sediment. The diesel is then returned to the diesel tank through a separator. Visit www.marine16.co.uk.

ABOUT USING FUEL ADDITIVES

The chemical makeup of fuel will designate the performance characteristics. There is no perfect fuel, so it is a compromise. Additives have many compounds to improve fuel quality, which can improve ignition delay periods and increase combustion efficiency. Additives also have detergents that keep injectors clean and improve lubricity. Anticorrosive elements are also incorporated to protect the injection system, and antifoam compounds are used to limit frothing; some have biocides to prevent microbial activity. There are many additives on the market, and I have used several, including Penrite and Sta-bil. Other products are Biobor JF, Starbright Star-Tron, Bio Kleen, Stanadyne, and several others.

ABOUT FUEL SYSTEM BLEEDING

When any part of the fuel system is disconnected during maintenance or troubleshooting, or air has entered the system, the fuel system must have all air expelled. Air within the fuel system can cause many problems and can be difficult to expel. Air in the system will absorb the pump force and the injector will not open. The injector pump and injector operate by having a high enough hydraulic pressure applied to open the needle valves. This allows fuel to enter under sufficient pressure to atomize for combustion. Do not attempt to start the engine until the injection pump is properly filled and primed or damage can occur, as fuel acts as a lubricant for the fuel pump.

SOME USEFUL FUEL BLEEDING ADVICE

1. Paint the bleed screws with bright yellow paint to improve visibility in low light.
2. Always have the correct or even dedicated wrenches and spanners ready by the engine for use.
3. Use caution with high-pressure fluids such as fuel, as escaping pressurized fuel can penetrate the skin, causing serious injury such as gangrene. If high-pressure diesel fuel oil is injected into skin, it must be surgically removed within hours.
4. Tighten all connections before pressurizing the system.

HOW TO BLEED YOUR FUEL SYSTEM

The bleeding sequence is generally as follows, but you should always consult your specific engine manual for applicable procedures:

1. Check that the diesel fuel tank has enough fuel.
2. Check that all fuel system pipes and connections are tight.
3. Check all hoses, fuel line fittings, and steel lines for damage, as they may chafe during engine operation.
4. Loosen the bleed screw located on top of the engine fuel pre-filter or separator, if one is installed.
5. Manually operate the thumb or hand-priming lever located on the lift pump. This must be done until a steady stream of fuel flows. Initially there will be a lot of bubbles; bleed until fuel flows bubble free.
6. When finished, tighten the bleed screw.
7. Repeat the bleeding process at the fuel injection pump inlet connection.
8. Loosen the bleed screws on the injection pump housing. This should be around two to three turns. Manually operate the lift pump until fuel is flowing without any bubbles.
9. Each injector must be bled and tightened in sequence. Start with the injector located at the end of the system from the inlet first. Tighten each in turn. Repeat this for each of the other injectors. Crack open the connections to allow air to bleed out.
10. Start the engine; if it operates normally, the remaining injectors should bleed automatically.

FUEL SYSTEM TROUBLESHOOTING

1. A loose fuel line on the suction side can cause ingestion of air into the system, resulting in a low fuel pressure problem.
2. A fuel pump can be damaged internally by a seized plunger or some other fault, such as a ruptured diaphragm.
3. The fuel filter can be clogged and cause a restriction in fuel supply, causing low pressure, so check the filter first. Note that after filter changes, you frequently have to bleed air out of the system by cranking over or running the engine.
4. You can get debris within a fuel regulator valve if installed; cleaning is again the answer.
5. A kinked or crushed flexible fuel supply line can restrict fuel flow and cause low fuel oil pressure.
6. Loose fuel oil suction line connections can contribute to air ingestion.

ENGINE SPACE FUEL SAFETY

Standards include the EI 15 Energy Industry requirement for oil lines above 1-bar pressure, and this is focused on oil mist prevention. The International Convention for Safety of Life at Sea (SOLAS) is a marine industry regulation that includes requirements for shielding and screening of oil spray within vessel engine room and compartments. The vast majority of engine room or compartment fires are initiated by an oil leak and subsequent ignition of the oil mist. Oil leaks are caused by small perforations of fuel lines; failure of a flange, connection, or valve; or a fracture or leak of pressurized oil, which can be diesel, lubricating oil, or hydraulic oil. When an oil leak occurs, oil mist is projected into the space. Leaks have sources in injectors or engine high-pressure fuel systems and oil lines that atomize the fluid when it escapes. Initially such escapes are invisible to the eye. Oil mist can be very small droplets, as small as 1 to 10 microns, and disperse within the surrounding air. When that oil mist attains the lower explosion limit and comes into contact with a heat source of 390°F (200°C), an explosion and fierce fire are often the result.

ENGINE SPACE FIRE IGNITION SOURCES

Ignition sources have diverse origins that include bearings, turbochargers, exhaust systems, and electrical sources such as electric contacts, faulty wiring, motors, and static electricity. There are things you can do to mitigate the risk; while this is an issue in larger sailing yachts, do a basic risk assessment regardless of your boat's size. The commercial maritime and offshore oil industries install what are called pipe and flange safety shields. Pipe safety spray shields are installed on boats to protect people from injuries and damage, fire, and explosions arising from high-pressure oil leaks and atomized oil mist. This hazard is created by spray-outs of fuel oil and hydraulic oil at failing pipe connections, including flanges, valves, and expansion joints. These spray shields are commonly known by the brand name FlangeGuard and come in a variety of materials. The original bag-type safety shield is a very cost-effective way to protect flange joints. The shield is simply wrapped around the outside of the flange and the protective fabric drawn down around the bolts to the pipe. The bag has drawstrings to hold the shield tightly in place. This design type is able to control the oil leak and prevent both oil spray-outs and the formation of oil mist. The escaping fluid coalesces inside the shield and then drains vertically in a drip or stream. The basic design has been improved by incorporating an outer band with specially formulated, multilayered internal mesh. This mesh is held tight against the flange and is the basis of what is termed pressure diffusion technology (PDT). The pressure is diffused at a controlled rate within the mesh. This is self-draining and prevents spray-outs and mist or vapor formation. There are valve protection options that fit butterfly, globe, check, gate, and bull valves. Check https://klinger-thermoseal.com for FlangeGuard details. Another product is Spray Stop Anti-Splashing Tape. It meets SOLAS requirements and is approved by all ship classification societies. Check it out at https://t-iss.com. Do a survey and a risk assessment on your own boat.

3

The Cooling Water System

Marine diesel engines universally have a heat exchanger with a closed-circuit freshwater (FW) cooling system. The FW system has a pump that circulates the water from the expansion tank, through the various engine water galleries, to carry engine heat away, through the cooler and then back to the expansion tank. The coolant system controls the overall operating temperature of the engine in operation, and proper heat transfer is essential. The freshwater cooling system must remain free of saltwater contamination to prevent corrosion or the formation of sludge and scale that may impede coolant flow or block coolers. This reduces heat transfer rates by coating engine block water passages with insulating scale buildup. This will result in gradual overheating, with all the damage that comes from it. It is important that inhibitors are maintained at the correct concentrations if performance is to be maintained and damage avoided.

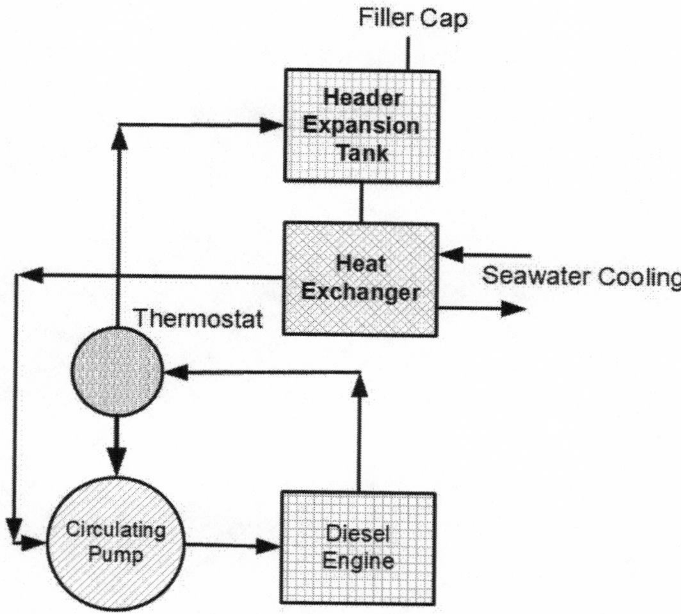

Figure 3-1. Engine Freshwater Cooling System

ABOUT COOLANT ADDITIVES

A number of additives are available to improve the performance of coolants. Coolant water may contain sulfates, chlorides, dissolved solids, and calcium. Coolant should have an antifreeze additive to prevent freezing and engine damage in cold climates. Most ethylene glycol–based antifreeze solutions contain the required inhibitors for normal operation. Check your engine manual for specific recommendations.

ABOUT CORROSION INHIBITORS

These are generally water-soluble chemical compounds that protect the metallic surfaces within the system against corrosion. Compounds can include borates, chromates, and nitrites. Soluble oil should not be used as a corrosion inhibitor.

ABOUT ENGINE COOLING SYSTEMS

There are three ways of cooling a marine diesel engine: direct seawater (SW) cooling, indirect or inter-cooling, and keel cooling.

COOLING WATER SYSTEM CLEANERS

One well-known cooling water cleaner is Sea Flush, which is a patented formulation for cleaning and flushing engine, generator, and air-conditioning heat exchangers, oil coolers, and other system components. Descalers are usually caustic, so care must be taken.

ABOUT DIRECT COOLING

Raw seawater is drawn directly through the hull scoop and grating, seacock, and a strainer. The raw water pump circulates the seawater through the engine block to cool it, after which it either exits straight overboard in a dry exhaust or through the exhaust line in a wet exhaust installation. The engine will have a thermostat to bypass the cooling water until the engine warms up. These engine types are no longer made.

ABOUT INDIRECT COOLING

The indirect or inter-cooled engine has an SW system and an FW system. The engine FW system is a closed-loop system that uses a pump to circulate the water from the expansion tank, through the various engine water galleries to carry heat away, back through the cooler or heat exchanger, and then back to the expansion tank. The heat exchanger provides the interchange for transferring engine heat from the FW to the SW coolant. The seawater is drawn from outside via the water scoop and grating and through the SW inlet valve, the strainer, to the suction side of the SW pump. The pressurized water is then pumped to the cooler, where it passes through the cooler tubes and transfers heat from the freshwater. The seawater then may be injected into the exhaust line and discharged overboard through the exhaust outlet.

ABOUT KEEL COOLING

A keel cooling system is similar to the indirect cooling system except that the external seawater system is replaced with a closed-loop system. This comprises a hull-mounted radiator that allows heat transfer from the circulating freshwater to the surrounding seawater.

ENGINE SEAWATER COOLING SYSTEM

The heat exchanger provides the medium for transferring engine heat from the FW to the primary SW coolant. The flexible impeller blades create close to a perfect vacuum. As the impeller rotates, each successive impeller blade draws in seawater and transfers it from the inlet to the outlet port. The pressurized water is then pumped to the heat exchanger or cooler, where it passes through the cooler tubes and transfers heat from the freshwater. Seawater cooling circuits are dependent on regular maintenance, and coolers require regular inspection and cleaning. Do not neglect pump impeller inspections, do change them every season, and do carry a spare plus a removal tool if required.

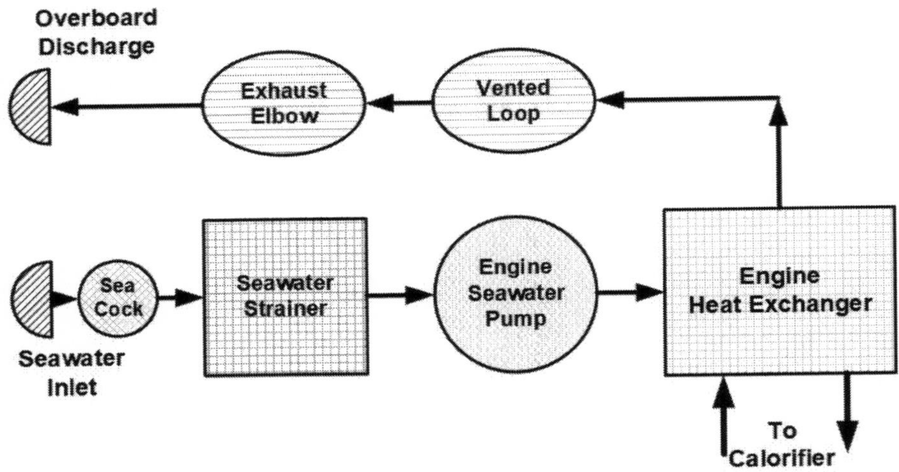

Figure 3-2. Engine Seawater Cooling System

SEAWATER (RAW WATER) STRAINERS

The raw water strainer is crucial to efficient engine cooling. It is frequently fouled and partially blocked with everything from jellyfish, rags and an ever-increasing amount of plastic, the remains of fish, and assorted detritus. Strainers need inspection and, if necessary, cleaning weekly when out on the water. They are installed in the seawater suction line above the waterline so that removal doesn't end up flooding the boat. It is always good practice to close the seacock before servicing, but don't forget to reopen it. The strainer housings have a clear plastic cover to enable easy inspection. The cover comes off and the strainer cage or element then lifts out for cleaning. The top cover has a seal that consists of a nitrile O-ring or gasket. Always tighten the lid properly to prevent air being sucked in by the pump.

STRAINER MAINTENANCE

Strainers need to be cleaned weekly when out on the water, and very muddy conditions can quickly result in strainer basket clogging when combined with other material. Strainers are required to continue allowing water flow as they become clogged, and the pressure drop between 0.1bar and 1bar can be as much as a 60% water flow decrease. Wear gloves when cleaning strainers—you don't know what is in there. I have been stung by various jellyfish when putting my bare hands into strainers. Strainers have a transparent polycarbonate top or sight glass to allow inspection. A common problem is overtightening the fastening nuts, so use care when reinstalling. If the strainer has a screw-on plastic top, don't overtighten and distort the seals. Check the seals for deterioration. Groco recommends lubricating cap threads and O-rings yearly with a Teflon or silicon grease.

ABOUT WATER INTAKE SCOOPS

The water intake scoop is designed so that when under way, the water is forced into the water intake, creating positive pressure. This scoop is a cast fitting that has a filter or strainer over it to prevent ingress of weeds and other material. The scoop is always installed facing forward in a motorboat. In a sailing vessel the scoop is installed facing to the rear so that water is not forced back into the system when sailing. If a water pump impeller is leaking, water can pressurize through, filling the exhaust line and ultimately back to the engine internal exhaust. The strainers on the scoop are prone to fouling with material and marine growth, and it is good practice to examine and clean them regularly to avoid starving the engine of cooling water and then overheating. This is very common, and the biggest enemy is plastic bags or parts of them. When painting or applying antifoulants, make sure you don't coat the grills too thickly or reduce the opening. Check your engine manual for guidance on installation.

ABOUT THE SYSTEM HOSES AND CLAMPS

Seawater inlet hoses should be of the wire-reinforced type. The flexible piping installation must avoid any excessive bend radius or kinking. Always install *two* stainless steel hose clamps at each connection and tighten properly. Piping can collapse and cause intermittent overheating. This often happens at higher running speeds and increased engine compartment temperatures. It is good practice to regularly examine hoses for hardening or pinhole leaks.

ABOUT THE WATER PUMP IMPELLER

Many impeller pumps are of the Jabsco type and are powered using a drive belt driven off the crankshaft pulley. When temperatures are high, the impeller may become fatigued. Follow this simple change procedure. Remove the pump cover, making sure not to damage the gasket. Carefully pry out the impeller with a screwdriver (use a special tool if it is recommended). Check the impeller for damage, cracks, and flexibility. Replace or renew, then coat with water pump grease only; do not use petroleum jelly (Vaseline) as this may degrade the impeller. Refit the gasket and pump cover; do not overtighten the screws. Make sure you have the pump details recorded, including the model and serial number, so you have the correct spare on hand.

WATER PUMP IMPELLER TROUBLESHOOTING

If the impeller is damaged, the following are possible causes that should be investigated and resolved:

1. Check if there are pieces missing out of the blade tips at the center of the impeller, pitting at the ends, or the edges have a hollowed-out appearance. This is caused by cavitation due to low pressures at the pump inlet. It can be rectified by reducing inlet pipe restrictions and lengths, and increasing inlet pipe diameters.

2. Check if the impeller blade tips and end faces are worn, or if the impeller drive is worn. This is caused by cavitation from low pressures at the pump inlet, and the same corrective measures apply.

3. Check if the end faces of the impeller have a hard and polished appearance, or if some or all blades are missing. This is caused by running the pump without water. The pump should not be run longer than 30 seconds without fluids, and should be stopped as soon as the fluids are gone.

4. Check if the impeller blades have excessive or permanent distortion or curving. This is caused by chemical action, excessive pump storage periods, or the end of normal service life. Chemical actions are caused by pumping incorrect fluids. If an engine is stored for long periods or over winter, remove the impeller to avoid damage.

5. Check if the impeller binds inside the pump housing, or if the blades appear longer than the hub. Check if the impeller rubber is sticky and soft—this is caused by chemical actions, high fluid temperatures, or long immersion. Pumps should be flushed clean after use and drained if being stored. High fluid temperatures should not be used.

6. Check if the impeller blades are cracked by 50% or more, and if parts of the blade are missing. This is caused by the impeller reaching the end of normal operational life, perhaps due to high output pressures. High- or low-temperature fluids and running the pump dry can cause similar damage. Check and reduce the pump pressures and outlet pipe restrictions, such as long pipe runs or blockages.

ABOUT SEACOCKS AND THRU-HULLS

The seacock and its attached hoses are all that separates you from being a floating vessel or a reef. The seacock is innocuous, but when you unpack the elements of watertight integrity, things change. Most boats have several seacocks to provide seawater to propulsion engines and generators, for toilet seawater flushing, for air-conditioning and refrigeration cooling, and more, along with an overboard discharge for all of these. Refer to American Boat and Yacht Council (ABYC) Standard H-27. Apart from ABYC, there are other recommendations. Standard ISO 9093:2020 requires that seacock and thru-hull fitting components formed of a metallic material show no degradation to the point that their operation is impaired. This ISO standard specifies requirements for thru-hull fittings, seacocks, hose connections, fittings, and installation in small craft.

ABOUT SEACOCK TYPES

The standard for a very long time has been either bronze or what is known as dezincification-resistant brass (DZR) seacocks. The metallurgical properties comprise a mix of copper, zinc, tin, and so on to increase the physical strength. Bronze is an alloy of copper, tin, nickel, and various other metals. The Blakes & Taylor (B&T) seacock has a tapered core plug assembly that facilitates valve adjustment without removal; it is easy to disassemble and service. Having had a B&T seacock on a previous old wooden boat, you do need to annually disassemble it, clean it thoroughly, lubricate it correctly, and lap it in if necessary. The seacock was over fifty years old and still leak-free and suitable for use. Similar bronze fittings are manufactured by both Perko and Groco; however, they are not the conical type, but use ball valves made from stainless steel or a polymer. This has been joined by the Forespar Marelon valves. These are not PVC or plastic, as they are often disparagingly described. They are a formulation of polymer composite compounds that use composite-reinforced DuPont Zytel polymer and additives or glass-reinforced DuPont Zytel; they are ABYC recognized, Underwriters Laboratories (UL) approved, and International Standards Organization (ISO) certified. Other composite seacocks are those from TruDesign, which have a Teflon-impregnated ball.

HOW TO PERFORM SEACOCK MAINTENANCE

The prime cause of seacock failures of all types is a lack of basic maintenance. If you don't perform the maintenance, you threaten the seaworthiness of your vessel and the safety of everyone on board. Always have wooden bungs tied close to the valve for emergencies.

1. **Winterization.** Seacocks are very prone to freeze damage in some locations, and they should not be left full of water. Prepping for winter entails closure, and draining water through the drain plug. Once drained, reinsert the drain plug.

2. **Lubrication.** Marine growth does occur within the valve but you can reduce this. Many seacocks have a grease nipple or zerk fitting. You should lubricate according to the manufacturer's recommendations. Groco recommends its own lubricant, Groco U-LUBE, a specially formulated grease for ball valves that inhibits marine growth and is injected into the space between ball and valve body. The company recommends opening the valves when greasing and removing the drain plug. Then charge the valve with grease; when grease appears at the drain plug, stop and reinsert the plug. In most cases the hose and hose clamps will require removal to access the valve to lubricate it. Using winch grease or a product called LanoCote grease is very effective. Water pump lubricants are also okay. There are polytetrafluoroethylene (PTFE) lubricants that do not attract and hold dust or grease. PTFE (Teflon) lubricant is a low-molecular lubricant formulated for applications that require a low coefficient of friction, good surface adhesion, and corrosion resistance, which is exactly what a seacock requires.

3. **Valve Exercise.** Regular maintenance requires monthly operation of the valve, repeating five times each time you do this. The valve should close without requiring excessive force or offering much resistance. Otherwise valve servicing or investigation is required. If a valve is installed correctly, you should be able to operate the valve handle through 90 degrees, allowing easy identification of status, whether open or closed. It is prudent to label the valve open and close positions for crew members who are less aware, to avoid damage.

4. **Corrosion.** Visually inspect for evidence of corrosion or dezincification on both the body and the hose tail, as well as any other attached fittings. *Never mix various fittings of differing metal composition*. I have encountered bronze valves with brass adapters and fittings; no doubt, the easiest way to source the parts was from the local hardware outlet. However, the use of different metals is a source of galvanic corrosion, and a bronze valve with brass attachments will not last long. Inspect the valves every year when hauled out for signs of dezincification. Gently scrape the surface; if it looks pinkish or is acquiring a dark red color, it is time to replace it, as the zinc component of the alloy is leaching out.

5. **Hose Tails.** The hose tail, spigot, tail pipe, or pipe to hose adapter is a male fitting that connects the hose to the body of the valve. The hose tail has barbs that hold the hose in place by friction. Hose clamps are used to secure the hose in place. Installing hoses and removing hoses to hose tails is always a challenge. You can put the end in hot water or gently use a hair dryer to heat up and soften the hose end.

6. **Hoses.** Use the right hoses, and spend the money to get the best. For cooling fluids, hoses manufactured from ethylene propylene diene monomer (EPDM) rubber that incorporate synthetic fabrics along with spiral steel reinforcement are ideal. These are suitable for cooling water and have a wide range of temperature capabilities. These hoses will not fold shut or kink, which often happens on seawater inlets and creates many overheating incidents.

7. **Hose Ends.** Hose ends are another variable to this equation. PVC hoses are often used, as are heavy-walled, wire-reinforced, or other types of reinforced hoses. The hose diameter should be the same size as the hose pipe. I have frequently come across people trying to clamp on an oversized hose or attempting to push an undersized pipe onto the hose pipe. Before installing Jubilee clips or worm-dry clamps, the hose ends should be clean and fresh. Do not install any hose end that is aged, brittle, or fatigued. Hose ends deteriorate because they are partially stretched over the barbed hose tail. The metal barb and the valve body suffer from cycles of hot and cold temperatures, which contribute to material hardening and degradation. When installing a hose, always allow some slack for trimming after each hose removal and reinstallation. As many know, removing hose ends is very difficult; it is usually quicker and easier to simply cut it off and remove from the barbed hose end.

8. **Hose Clamps.** The humble hose clamp, Jubilee clip, or worm-drive clip has a lot of responsibility. These clamps are also called pipe clips, hose band clamps, or hose locks. It is an essential part of the thru-hull and seacock arrangement, along with the hoses that connect to it. Many install the wrong grade of clamps, such as plated steel or 304 grade stainless steel. Always select 316 grade stainless steel, and check the clip for markings. They are used on seawater intakes and discharges and on freshwater, sewage, and bilge pump overboard lines, with many systems being under pressure. If they fail, you can end up with a very big mess or, worst case, your boat sinks—all that riding on the integrity of a hose clamp.

9. **Hose Clamp Installation.** Use the correct clamp for the hose; if it is too large, you will be unable to apply pressure evenly and the tail may be damaged. Long tails are a danger, and I have frequently been lacerated from sharp stainless edges, once requiring an emergency room visit. You can buy small plastic hose clamp end guards or protectors from Clamp-Aid; given my past painful experiences, I absolutely recommend them. You can use color-coded ones to mark your seawater and freshwater systems and so on, which is a bonus. The traditional Jubilee worm-drive clip is identifiable by the short and thick worm drive, which has an integral grub screw or set screw mechanism located on the band top.

10. **Hose Clamp Tightening.** When you tighten clamps, they pull in and grip the band circumference on one side for maximum band tension and clamping force. Most clamps are tightened using a large screwdriver blade, a small socket spanner, or a hex key. Personally, I use a ratchet drive and socket, which is more effective than a screwdriver or hex key to torque properly in confined work areas. Do not overtighten, as you may end up damaging the clamp threads and the hose. Another alternative is the Super Clamp from Jubilee. These allow torquing to 18ft-lb (25Nm) and incorporate 8.8 high-tensile bolts; they are specifically designed for suction and delivery hoses where a tight seal is required. Bolt drive or T-bolt versions have a thinner profile and are also known as T-bolt clamps or head bolt clips. They are distinguished by the small bolt located across the band top. When the bolt is tightened with a hex drive, it pulls the hose clip in from both directions and evenly applies the torque.

11. **Hose Failure Modes.** The most common failure mode is the distortion or breaking off of valve stems or handles, or shearing the stem completely. People try to force them the wrong way or force them against the stop, rendering the valves inoperable. The valves are often seized, or people simply do not know how to operate a valve. When operating a valve, if it is hard or stiff to move and requires force, you need to service the valve. Another key failure mode is stepping on or accidentally applying force to the attached hoses. Often seacocks are installed in areas prone to damage. Stepping on a hose creates what is known as the lever-arm effect. The stiffer and larger the hose, the more prone it is to damage. Another failure mode is when people try to pull off hoses from barbed hose tails and end up damaging the valve through excessive force.

THE HEADER OR EXPANSION TANK

When water is heated, it expands; when cooled, it contracts. The expansion tank allows this to happen, and the tank should be kept at the level recommended for your engine. Remember to check frequently and never open the cap when hot.

ABOUT THE FRESHWATER PUMP

The freshwater pump is critical to the performance of the cooling system. It is usually a centrifugal pump belt driven off the engine and is relatively reliable. The pump impeller is made of neoprene rubber. Again, monitor the weep or telltale holes for leakages. Regular inspection and programmed replacement are recommended. Check the rubber V-belts. They are not just for alternators, and excessive tension can cause bearing failures in pumps.

THERMOSTAT OPERATION

This is essentially a heat-activated valve. It restricts the circulation of the engine coolant until the engine heats up to the nominal temperature setting. When the engine heats up, the thermostat opens to allow more water to circulate through the heat exchanger. A thermostat starts to open around 160°F (70°C) and is fully open at 185°F (85°C)

ABOUT THE HEAT EXCHANGER OR COOLER

The heat exchanger transfers engine heat from the freshwater cooling circuit to the raw seawater circuit. The cooler tubes are prone to blockage through silt and scaling, so in high silt water areas regular inspection is recommended. Scaling can be reduced by idling the engine for 10 minutes to cool it down before stopping. I once had a high engine temperature issue and discovered that the cooler tubes were caked in fine silt from the brown, heavily silted river we were transiting in Europe.

ABOUT COOLER ZINCS OR ANODES

Sacrificial anode or zinc pencils are used to counter the effects of electrolytic corrosion in heat exchangers and cooler casings. These cooling water elements also are subject to galvanic corrosion due to the differing metals used within the system. They should be removed and inspected, and replaced every 200 hours.

HOW TO INSPECT AND CLEAN COOLERS

The following essential maintenance tasks should be completed regularly to maintain optimum performance. When temperatures cannot be maintained, coolers need to be opened, inspected, and cleaned. This task requires a soft wire brush and a soft metal rod (copper or brass) that is slightly smaller than the tubes:

1. Close any isolation valves.
2. Drain the cooler.
3. Remove the cooler end bonnets.
4. Clean all accessible parts with the soft wire brush.
5. Push the rod through each tube. In severe cases of fouling, acid cleaning may be required.
6. Flush the cooler and tubes well with fresh water.
7. Inspect and renew the gaskets and seals if perished or damaged.
8. Reassemble the cooler.
9. Run the engine up to normal speed and observe oil and water temperatures. Inspect for leaks.

ABOUT COOLANT ADDITIVES

Coolant additives have a primary function of providing freeze protection. The ethylene glycol–based solutions, commonly referred to as inorganic acid technology (IAT), provide freeze protection down to about -30°F (-34°C) and raise the boiling point of the cooling water by approximately 15°F (-9.4°C) above nominal water. Coolant water must be chemically pure and should be either deionized or distilled and demineralized. Additives are used to lubricate the water pump impeller and protect against cylinder liner erosion and pitting and boilover, and to create a stable and noncorrosive system environment. This will offer protection and long life for internal metal parts of the system, hoses, gaskets, and seals. Corrosion inhibitors are water-soluble chemical compounds that protect the metallic surfaces within the cooling water system against corrosion. Compounds can include borates, chromates, and nitrites. Additives can reduce coolant vapor bubble formation by depositing a protective film on the cylinder liner surfaces that acts as a barrier against collapsing vapor bubbles. Soluble oil should not be used as a corrosion inhibitor. Only use non-chromate inhibitors. Maximum levels allowed are chlorides 40ppm, sulfates 100ppm, total dissolved solids 350ppm, total hardness 170ppm, and Ph level 5.5 to 9.0. They can be checked using test strips every 600 hours of operation. Do not use general automotive-type coolant additives, as some may contain high silicate concentrations that can damage marine diesel engines. Consult your engine manual for the correct coolant additive recommendations.

4

The Lubricating Oil System

Lubricating oil separates the various engine working surfaces to prevent metal-to-metal contact. The goal of lubrication is to reduce friction by substituting fluid friction for sliding friction. Oil assists in bearing cooling and prevents corrosion. The pistons must receive adequate oil supplies to prevent expansion and seizure that is caused by excess heat due to increased friction on a dry cylinder wall. Lubrication has the dual function of lubricating the engine's moving parts and removing heat generated during the combustion process and friction, which reduces wear. Oil removes heat from the outside of cylinder liners and from the pistons and inner walls of liners, and cools the main bearings. Friction can create enough heat to melt bearing metals.

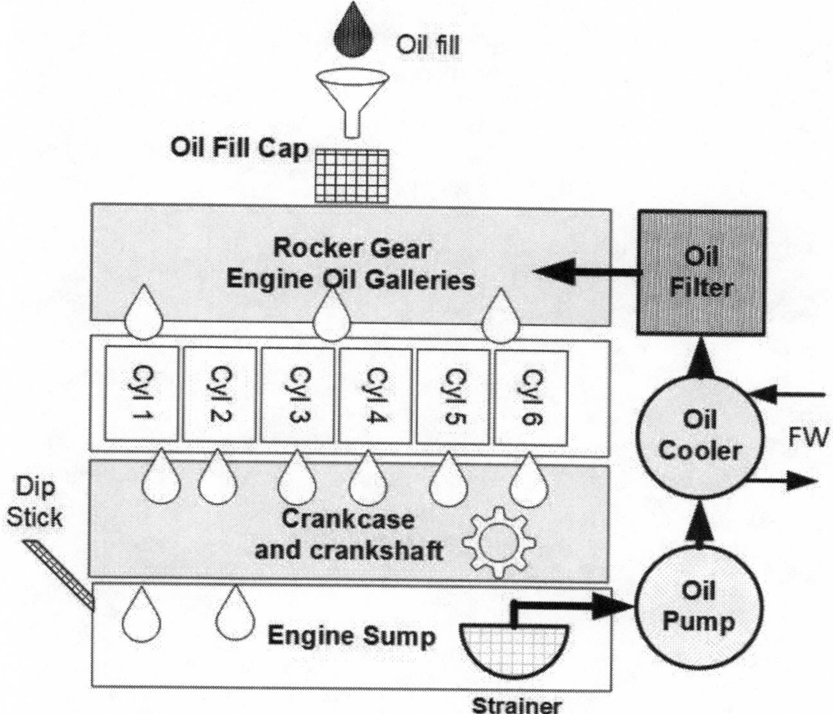

Figure 4-1. Engine Lubrication Oil System

ABOUT LUBRICATION OIL CIRCULATION

The lubrication oil is taken to the oil pump from a submerged suction in the oil sump or pan. Oil pumps are usually a two-stage, positive displacement, gear-type pump. Oil pumps are driven off the timing gear train to maintain constant oil flow. It pumps a constant volume of oil, although flow will decrease slightly with increasing pressure. A nonadjustable pressure regulator valve is installed on the pump outlet side. A simple bypass valve is used to limit maximum oil pressure by diverting excess oil back to the sump. The pressurized oil is passed though the oil filter then distributed around the various oil galleries and points within the engine. As oil reaches the crankcase bearings, flow is restricted, and the pump forces oil into the clearances between the main bearings and crankshaft. Oil is carried to the crank, connecting rod bearings, the main journal, and connecting rod journals. Oil may be used in turbo bearings and sometimes under piston crowns. Liners, pistons, and rings are lubricated by oil thrown by the camshaft and crankshaft, and the oil manifold supplies oil to the governor and rocker gear. Oil usually flows off the end of the rockers' arms to splash-lubricate valve springs and stems. Oil also lubricates the timing gear bearings. The oil returns back to the sump via gravity. In many engines the heat is dissipated through the oil cooler and through oil sump surfaces.

ABOUT OIL FILTRATION

All oils contain contaminants, and the content levels must be minimized. These may be metals, fibers, or microbial growth, and most damage is caused by hard particles that are slightly larger than the clearance of contact surfaces. Critical sizes are 0.5 to 450 microns and usually greater than 3 to 5 microns. The oil filters remove these abrasive materials and contaminants. Oil protects against corrosion and helps the piston rings seal the cylinders and reduce compression loss through the rings. Oil provides a strong film between all surfaces of working engine parts so that wear is reduced. Oil films may be typically around 30thou (0.08mm) thick between the shaft and bearings, and must be strong enough to prevent surface contact.

ABOUT OIL CONTAMINATION

In a four-stroke diesel engine, lubricating oil reduces engine wear and absorbs contaminants that enter the engine. Most contaminants are expelled from the engine with exhaust gases, but some remain in the cylinders and crankcase to corrode metal parts and form sludge and lacquer deposits. Most events that occur in the engine have an effect on the oil. When injection and combustion pressures are raised, the loads increase on pistons and bearings. Excess oil will bypass the pistons and rings into the combustion chamber, where it burns, forming carbon deposits. Oil is affected by chemical composition and the presence of nitrogen, phosphor, potassium, sulfur, and external growth factors such as temperature, pH values, water, and oxygen. Oil neutralizes the acids that form and breaks up the deposits caused by combustion blowby. Oil cleans by dissolving sulfur that is converted to acid when the fuel is burned. High oil temperatures cause chemical breakdown of oil, so maintaining the correct engine temperature is essential. Reduce the effect of acid formation by operating the engine at proper working temperatures. Humidity in combustion air also assists in acid formation. When sulfur is low, at around 0.02%, there is little affect, but at 1%, humidity will affect acid formation. Oil at some point loses its effective viscosity as it transfers heat during combustion and absorbs combustion by-products. When oil completely oxidizes, it can lead to black sludge formation and subsequent engine damage. Regular oil changes are good for your engine.

ALL ABOUT OIL STANDARDS

It is essential that the correct grades of lubricating oil be used for the prevailing temperature conditions, and that the filter be changed regularly along with the oil. The nominal rating of oil viscosity must be maintained if correct lubrication is to be achieved. This is dependent on the engine remaining within the proper operating temperature ranges. Engine oil should comply with American Petroleum Institute (API) Service Classification API CC or API CD. These may be synthetic, mineral, or blended oils. Mineral oils have various additives and are suitable for infrequent use or under less difficult service, such as in marine applications. Oils in generators are changed at 500 hours; synthetic oils usually last longer by a factor of 10. This results in a cleaner engine, rings, and liners; increases engine life by 40%; and decreases lube oil consumption by 75%. Society of Automotive Engineers (SAE) standards match oil viscosity to operating temperatures. The "W" notation is the winter service rating, defined at 0°F (-18°C). Always consult your engine manual and manufacturer's guidelines for lubricating oil requirements.

OIL VISCOSITY EXPLAINED

A simple test method for viscosity uses a viscosity test stick or flostick. Both the new and used oil must be allowed to stand for an hour and stabilize at room temperature. The test stick is angled so that each oil sample runs down a channel. When the new oil reaches a mid-scale point on the graduated scale, the position of the used oil is read off. If the used oil has not reached the same point, it has a high viscosity, usually due to oxidation or high levels of insolubles. If the oil has run past the scale point, the additives may be failing or there may be fuel oil contamination. These test kits should be part of any testing regime, as they are simple and easy to perform. Viscosity tests measure a lubricant's ability to flow at a specific temperature. If the viscosity is reduced, an oil film cannot be properly established at the friction point. Improper viscosity leads to overheating, accelerated wear, and failure. This can occur due to heat and contamination not being removed at the correct rates.

Lubrication oils are classified by the ISO viscosity grade (VG). The ISO VG refers to a lubricating oil's kinematic viscosity at 104°F (40°C). If an oil's viscosity is within plus or minus 10% of its ISO grade, it is normal. When oil viscosity exceeds plus or minus 10% and less than plus or minus 20% percent, it is marginal. Viscosity that exceeds plus or minus 20% from grade is classified as critical. Low oil viscosity creates many engine problems, which is why good filtration and regular oil changes result in improved engine life and performance. Loss of oil film quality results in excessive wear due to increased friction on surfaces. The increased friction results in greater energy consumption and increased heat levels, as well as increased exposure to particle contamination due to reduced oil film levels. Thin oil films lead to high temperature at higher loads, or during engine startup. High viscosity leads to engine problems and is why you should gently warm up a very cold engine. It can cause excess heat generation, which results in oxidation of oil along with sludge formation. A condition called gaseous cavitation can occur, which is caused by inadequate oil flow to both pumps and bearings. High viscosity causes oil lubrication starvation due to inadequate flow and surface film creation. Poor air dissipation or demulsibility also can occur.

Engine Oil Viscosity Table

Viscosity	**Temperature °F**	**Temperature °C**
SAE 10W	-10°F to 70°F	-23°C to 21°C
SAE 30	+20°F to 100°F	-7°C to 38°C
SAE 40	+45°F to 120°F	7°C to 49°C
SAE 10W-30	-10°F to +100°F	-23°C to 38°C
SAE 15W-40	+5°F to 120°F	-15°C to 49°C

ABOUT FUEL IN OIL

Fuel in the lubricating oil can create a crankcase explosion risk and is characterized by low lube oil viscosity. Fuel mixing with lube oil reduces viscosity and reduces lube oil effectiveness, which increases wear due to a thinner film of oil on moving parts, including main crankshaft bearings. The main causes are faulty injectors, incomplete combustion, wear on engine parts such as valve guides and injectors, and cold starting.

ABOUT WATER IN OIL

Water in the oil can cause emulsification and degrade the oil's lubrication properties. Water causes internal corrosion, and dissolved water creates oxidation, which results in accelerated wear, increased friction, and higher operating temperatures. Water should not exceed 500 parts per million (ppm). The system must be completely flushed out after a leak repair so that no moisture remains. Common causes are external contamination from breathers and seals, along with internal leaks from heat exchangers and condensation.

HOW MICROBIAL GROWTH OCCURS

Where water and oxygen are present, microbial growth within the oil and system can occur. Anaerobic microorganisms can live in oxygen-free areas on partly mineralized hydrocarbons, and oil additives can stimulate growth of organisms. As the oil degrades, oil characteristics also change; this is caused not by the degradation but by the organisms that produce extracellular biopolymers (slime), usually bacteria and fungi, and yeast contamination. Organisms start to produce surfactants and biopolymers when in contact with what is called a hydrophobic culture medium (oil) that they can break down. This leads to increased exposure and oil breakdown. Optimal temperature range is around 50°F to 122°F (10°C to 50°C). Once the system is "infected," considerable flushing is required. There is a range of additives, but nothing surpasses good maintenance and monitoring.

ABOUT OIL ADDITIVES

Oil comprises hundreds of chemical compounds and often contains additives to improve the lubrication characteristics. Oil additives allow continuous operation at high speed or at raised engine temperatures. Oils have oxidation inhibitors and antioxidants to prevent oil thickening and formation of varnish and sludge that can seize fuel pump plungers. Antifoaming agents are used to minimize foam formation, which can cause a loss of oil pressure and a loss of lubrication. Air bubbles in oil will retain heat, and cooling ability will be greatly reduced. Detergents are used to prevent by-products of combustion from adhering to metal surfaces, as these deposits lead to excessive wear. Dispersal compounds such as calcium and barium prevent smaller particles within the oil agglomerating to form larger ones, which can cause damage by blocking oil galleries and passages. Dispersal compounds keep particles suspended in the oil, which can then be filtered out. Zinc and phosphorous act as lubricants to resist pressure inside the cylinder. Magnesium inhibits excess wear and corrosion. Additives deplete in normal engine operation and require renewal. Remember that the longer an oil is used, the greater the oxidation, which alters the viscosity. Alkaline additives are sometimes called buffers, and they prevent acid corrosion and wear on the internal engine parts. The alkalinity of oil is referred to as the total base number (TBN); the higher the TBN, the greater the ability of the oil to neutralize acids.

ABOUT LUBRICATION OIL TESTING

With larger vessels and engines, it is an economically sound investment to take oil samples at regular intervals and have an oil analysis performed. All leading oil companies offer this service, and analysis of the various trace metals found in the sample is a good indicator of engine wear. Different parts of an engine have different metals, which are all washed by lube oil. Oils are analyzed using a spectrograph. As each element is burned in an electric arc, the sample emits at a unique light frequency. The test results show amounts of each metal present in parts per million (ppm), and spectrochemical analysis is given in parts per million by weight. The analyst looks for wear metals, contaminants, and oil additives. Oil analysis should be done at regular intervals, as trend analysis is more important than a single test. Wear metals have differing thresholds, usually indicating ranges of increased probability that problems are developing. Sharp increases in wear metals or a major shift in physical properties signal impending problems.

ABOUT TAKING OIL SAMPLES

The oil sample must be taken uniformly, as wear metals are heavy and sink to the bottom of the oil sump if oil is settled and not circulating. Take the oil sample within 20 minutes of engine shutdown so that the oil material is equally dispersed in suspension. Do not sample from the sump bottom but from a point before the oil filter or suction, or out of the dipstick tube. Drop the suction hose to the sump bottom, then raise it an inch or two for the sample.

READING THE OIL TEST RESULTS

New engines will show higher wear metal numbers than old engines, and baselines are best established after the engine is broken in. On oil analysis reports, "normal" status indicates that the physical properties of the lubricant are within acceptable limits and no signs of excessive contamination or wear are present. A "monitor" status indicates that specific test results are outside acceptable ranges but not serious enough to confirm abnormal conditions. Initial abnormalities often indicate the same result patterns as temporary overloading or extended operations. An "abnormal" status indicates that lubricant physical properties, contaminations, or component wear are unsatisfactory but not critical. A "critical" status is serious enough to warrant immediate diagnostic and corrective action to prevent major long-term performance loss or in-service engine component failure.

ABOUT ENGINE WEAR METALS

There are typically more than forty different elements present in lubricating oil. Elemental spectroscopy is used to determine the concentration of wear metals, contaminant metals, and additive metals in a lubricant. Spectroscopy cannot measure particles larger than approximately 7 microns. Oil analysis results can indicate expected and unexpected elements. Some contaminates from within the engine are collected as the oil circulates, as different parts of an engine have different metals, all washed out by the lube oil. Other contaminates enter the machine from external sources, such as faulty seals and breathers, and during service and maintenance. Contamination usually presents in the form of insoluble materials. These include water, metals, dust particles, sand, and rubber. Even the very smallest particles, below 2 microns, are able to cause significant damage. Typically, these are silt, resin, or oxidation deposits. Elements that are indicators of contamination include silicon (i.e., dust and dirt or de-foamant additives), boron from coolant corrosion inhibitors, potassium from coolant additives, and sodium from detergent and coolant additives.

Engine Oil Analysis

Element	Indicators
Iron	High levels indicate wear from rings, shafts, gears, valve trains, cylinder walls, pistons, or liners
Chromium	May indicate excessive wear of chromed parts such as rings, liners, and some additives
Nickel	Secondary indicator of wear from some bearings, shafts, valves, and valve guides
Aluminum	Wear on pistons, rod bearings, and certain shaft types
Lead	An overlay on main rods and bearings
Copper	Wear from bearings, rocker arm bushings, pin bushings, thrust washers, and other brass-bronze parts
Tin	Wear from bearings and pistons in some engines
Silver	Wear from bearings; a secondary indicator of oil cooler problems, when coolant is detected
Titanium	Used as an alloy in steel for gears and bearings
Silicon	Airborne dust/dirt contamination indicates poor air cleaner servicing and can accelerate wear
Boron	A coolant additive and an additive in some oils
Sodium	A coolant additive and an additive in some oils
Potassium	A coolant additive
Molybdenum	Wear of rings; an additive in some oils
Phosphorus	Anti-rust agent and combustion chamber deposit reducer
Zinc	Antioxidant, corrosion inhibitor, anti-wear additive; detergent and extreme pressure additive
Calcium	A detergent, dispersant, and acid neutralizer
Barium	Corrosion inhibitor; detergent and rust inhibitor
Magnesium	Dispersant and detergent additive; alloying metal
Antimony	A bearing overlay alloy or oil additive
Vanadium	A heavy fuel contaminant

5

The Engine Air System

In a normally aspirated engine, the air for fuel combustion is drawn in through an air filter and compressed. The amount of fuel that can be burned, and therefore the power of the engine, is limited by the air mass within the cylinder. The options are to either pre-cool the air to increase the air density or use turbocharging. Air filter cleanliness is absolutely essential, and it is surprising how marine diesel engine air filters can get so dirty on the water. It doesn't take much filter blockage to cause power reductions. Make sure you carry a spare filter. Another enhancement to the combustion air system is the addition of a supply fan or blower. American Boat and Yacht Council (ABYC) Standard H32, Ventilation of Boats Using Diesel Fuel, has a formula. The normal heat load needs to be considered and compensated for with increased air supply. Not only does the engine receive a greater amount of air, but the whole space runs cooler, including rubber drive belts, alternators, and other parts.

ABOUT TURBOCHARGING

Turbocharging effectively increases the available power output for the same normally aspirated engine. Turbocharging increases the air density by increasing the air pressure when the cylinder is filled with air during the air intake stroke. This has the effect of increasing the engine power for the same cylinder size. The turbocharger is essentially a small air compressor driven by a turbine placed in the exhaust line. As the engine load increases, the hot exhaust gas output velocity also increases, which increases the turbine speed to drive the air compressor faster, which raises the air pressure into the cylinders. As the air is compressed into the engine cylinders, the air temperature increases, reducing available oxygen, so some engines may have intercoolers installed. These cool the compressed air, which improves combustion. Cooling air will increase the air density further, increasing the effectiveness of the turbocharger. The cooler is water-cooled to maximize heat transfer rates:

1. Cool air is drawn in through the engine air intake by the compressor fan.
2. The compressor then compresses the air and heats the incoming air.
3. The hot, compressed air is then blown into the air intake to the heat exchanger.
4. The heat exchanger cools down the air, which increases its density.
5. The cooler-compressed air then passes to the engine air intake.
6. The increased oxygen content of the air increases the combustion efficiency, resulting in improved energy production.
7. The exhaust gas from the engine then passes to the turbocharger turbine fan, rotating the turbocharger fan at high speed. These are mounted on the same shaft.

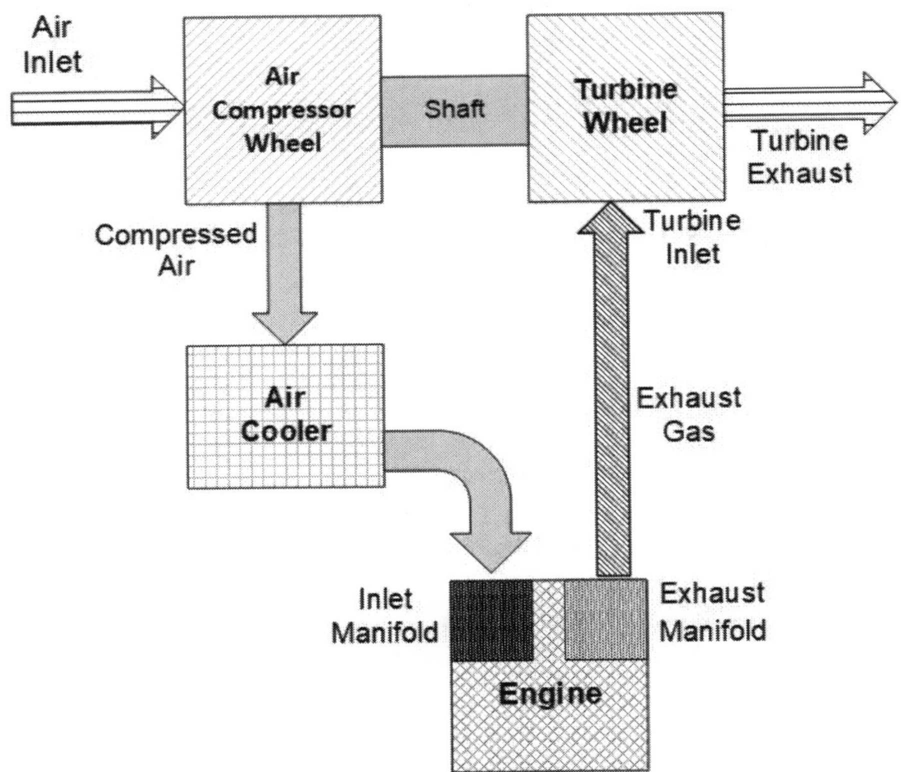

Figure 5-1. Engine Turbocharger System

TURBOCHARGER TROUBLESHOOTING

Turbocharger faults cause reductions in power output, generate black exhaust smoke, and increase oil consumption. The turbocharger shaft assembly should be inspected where possible to determine the fault. This requires the removal of the inlet and exhaust trunk. The turbine should be rotated by hand and the housing examined for signs of contact or rubbing. The oil drain should be checked and cleaned if fouled, and all oil leaks should be investigated. Engines subject to low speeds or extended idle periods tend to leak; this usually disappears when the engine is loaded up. The most common cause of turbocharger failure is hot shutdown. When an engine is shut down suddenly, the turbocharger continues to rotate without oil. Turbocharger service life is reduced due to bearing wear from inadequate lubrication. Eventually, bearing wear allows turbocharger casing-to-turbine contact and out-of-balance conditions. This can cause serious damage, or even destruction. When stopping an engine, it is good practice to operate at slow speeds for a few minutes. This allows the turbocharger to spool down and for cooling to take place. The engine should never be revved prior to shutdown, as the turbocharger will continue rotating without lubrication, which may damage the bearings. Turbochargers rotate at high speeds and generally use engine oil for lubrication, so maintaining clean oil is essential.

EXCESS TURBOCHARGER NOISE

Turbochargers tend to make serious noises when something is wrong or degraded. Possible causes are:

1. There are restricted or clogged air inlet filters.
2. The rotating turbine assembly is binding or touching the housing.
3. The flanges on the manifolds have become loose, creating leaks.
4. There is an object inside the compressor housing, inlet ducting, or manifold.

TURBINE ASSEMBLY IS BINDING

1. There has been ingress of material causing turbine or compressor damage.
2. The turbine or compressor wheel is contacting the housing due to bearing wear or failure.
3. There is an accumulation of carbon deposits in the turbine housing or on the turbine blades.

TURBOCHARGER SEAL LEAKAGE

1. There are restricted or clogged air inlet filters.
2. The oil drain lines are clogged.
3. The crankcase breather is clogged.
4. The bearings are worn or failing.
5. There is a piston ring leakage or high crankcase pressures.
6. The compressor wheel is damaged.

6

The Engine Exhaust System

Stringent emissions legislation is now being introduced and becoming mandatory in many parts of the world. It is inevitable that some fuel will remain unburned after combustion and come out in the exhaust. Maintenance is essential to reduce or keep this to a minimum. The exhaust engine noise is caused by the explosion taking place during combustion, and these shock waves resonate and echo through the cylinder and engine, then reverberate back out through the exhaust. The exhaust smoke color can be indicative of problems; normal exhaust has little to no color.

ABOUT WET EXHAUSTS

Exhaust gas temperatures can exceed 1100°F (600°C) on a diesel engine. When water is injected into the exhaust line, the high temperature is significantly reduced, and this also reduces both the pressure and volume. This lowers the exhaust gas speed, which results in a lower exhaust noise level. The temperature reduction allows the use of rubber compounds and a reduction in heat radiation to surrounding areas and materials. When the gases mix with the water, many of the particles within the exhaust are captured, making exhaust emissions much cleaner. In general, to ensure proper drainage of wet exhaust systems, the pipe from the injection point to the exhaust must be installed so there is a consistent downward angle along the entire exhaust length.

ABOUT THE INJECTION OR MIXING ELBOW

This is located at the exit point from the engine. In many cases, seawater is directly fed from the engine seawater supply into the elbow. There are cases where the water injection point is below the waterline, or less than 6in (1cm) above the waterline when the vessel is heeled. This creates a danger of the water siphoning back into the engine when the engine is stopped or when the boat is heeling. In this case it will be necessary to install an anti-siphon lock or air vent in the water line well above the waterline. The vent line, where installed, always goes to the transom, which is above the waterline.

A WARNING ON WET EXHAUSTS

Water locks are used to prevent water from going back to the engine. When a lot of starting attempts are made with a diesel engine, starting may become more difficult as water builds up within the water lock and silencer. Some water always remains from the previous operation. At each successive start attempt, additional water is injected as the engine turns over, without any exhaust pressure to blow it out. This excess water can be drained off using the small valve located at the bottom of the water lock.

ABOUT ANTI-SIPHON VENTS

This is also known as a vented loop, siphon break, or anti-siphon breaker. In addition to the anti-siphon note above, it deserves a separate section. This is a subject that requires careful attention, given that these valves are critical to the safety of the boat but are often never inspected or maintained, and are frequently almost inaccessible. Whether installed on the engine or generator exhaust system or on the toilet system, they need to be monitored and maintained. A valve malfunction can lead to your boat sinking. The vent is a single-direction, or one-way, breather valve that is installed at the top of the loop. The loop is installed as high as practicable above the dynamic or heeled waterline; recommendations vary between 1ft and 2ft (300mm and 600mm). The loop acts as a hill for water to flow up and over. The vent valve will allow air to enter the line when no water flow is following, which equalizes pressure and prevents the siphoning effect from initiating. When the engine is running and water is flowing up through the loop, the valve will close under water pressure and seal. The valve types vary from diaphragm types to duckbill or joker valves.

VENT INSPECTIONS

Inspection will require the removal of the vent installed at the top of the loop arch. Some disassembly is required, which is often challenging because they are in hard-to-access locations. Remove vents with care to avoid damage. Check the rubber, silicon, or neoprene membranes for deformities; if degraded, fatigued, or encrusted with salt and debris, it should be replaced. Other valve types incorporate spring-loaded flaps or balls, which should be cleaned and checked; in some cases, they may require lubrication. Vetus recommends that the valve assembly be coated with a Teflon spray or silicon oil before reassembly to ensure good operation.

SILENCERS AND MUFFLERS

This is usually a molded plastic chamber that allows exhaust gases to expand and have less volume, and therefore reduce exhaust noise. They are often molded so that they create a swirling action of the water to reduce exhaust noise. These should be mounted, as much as is practicable, in a vertical position. The installation of a silencer is important, and the manufacturer's instructions must be followed carefully. In general, the silencer should be installed as close to the midships position as possible to prevent water flowing back into the engine when the boat heels. The silencer inlet point must remain below the exhaust injection elbow at all times.

ABOUT WATER LOCK FUNCTION

In a wet exhaust, once the engine stops, there will be some water remaining in the exhaust line. Water must be prevented from draining back to the engine where it would cause serious damage. The water lock, sometimes called a water lift muffler or water trap, stops this flow by capturing and retaining it. When the engine next starts, the water is ejected (you will often see a sizable spurt of water at startup). The water lock is typically made from molded plastic, and is installed lower than the engine and exhaust outlet so that all water drains down to it.

ABOUT THE GOOSENECK

The gooseneck is often installed adjacent to the transom exhaust outlet. In some boat installations, the hose is looped up, with the top of the loop being at least 16in (40cm) above the waterline. In many smaller vessels, this clearance is not possible in the small space under a cockpit floor. The solution is to install a molded plastic gooseneck that prevents and captures any water ingress. The gooseneck has a secondary muffler function.

ABOUT EXHAUST HOSES AND CLAMPS

Hoses are made using a special multilayer rubber compound construction with steel spiral reinforcing. The exteriors have a rough appearance and are impervious to fuel, saltwater, and so on. Hoses are internally very smooth to reduce resistance and resultant back pressure. Approved hoses are rated at temperatures ranging -100°F (-73°C) to 212°F (100°C). Exhaust hoses should fit neatly onto the various components, such as the water lock and mufflers. Do not use oil or grease to slide them on; only use water or soap as a lubricant. Always use *two* stainless steel clamps on every connection, and ensure they are tight. Always properly secure or support all parts of the exhaust system, including the hoses, as they will contain the added weight of water within them.

ABOUT THE TRANSOM EXHAUST OUTLET FLAP

The flap is a hinged assembly that is closed when the engine is off and opens under exhaust gas pressure. This prevents water from following seas surging back down into the exhaust line.

DIESEL EXHAUST SMOKE

In normal operation, well-maintained diesel engines do not create smoke, and exhaust gases are clear. Monitor exhaust at all times; when the color changes, investigation and action are required. The mantra for a properly performing diesel engine is given as "clean fuel, clean air, and clean lubrication oil." While it is good practice to allow a diesel engine to warm up, when you have finished a run, allow a few minutes for the engine to cool down slowly.

WHITE OR GRAY EXHAUST SMOKE

If you observe white exhaust smoke, it is possible you have fuel injection problems and incomplete fuel combustion. This may happen when an engine is started at very low temperatures but soon disappears as the engine warms through. Causes include diesel fuel passing through unburned, coolant entering the combustion chamber, or the engine temperature being too low. A thick white cloud is serious, and the engine should be stopped immediately. Diesel engines require precise injector pump timing and fuel delivery at the correct pressure. A decrease in either results in white smoke from incomplete combustion:

1. Low cylinder compression: broken or leaking valves.
2. Low cylinder compression: worn piston, piston rings, and cylinder.
3. Low cylinder compression: stuck piston ring.
4. Low cylinder compression: leaking or blown head gasket.
5. Low cylinder compression: liner glazing.
6. Low cylinder compression: cylinder head or block cracked.
7. Injector fault: stuck open (timing gear worn).
8. Injection timing problem.
9. Fuel pump low pressure: worn pump.
10. Fuel pump timing problem.
11. Clogged fuel filter.
12. Glow plug problems.
13. Coolant water mixed with diesel fuel: cracked head.
14. Cylinder head gasket failure: coolant leak.
15. Cylinder head or block cracked: coolant leak.
16. Cylinder head damage: coolant leak.

BLACK EXHAUST SMOKE

Black exhaust smoke indicates combustion issues. It can be an indicator of too much fuel or not enough, or excess air or air starvation. Bad combustion is economically poor, environmentally damaging, and could herald major and expensive damage if not dealt with promptly:

1. Damaged turbocharger or intercooler.
2. Dirty or clogged air cleaner: air restriction.
3. Excessive engine sludge buildup.
4. Low compression and worn piston rings.
5. Dirty, worn, or malfunctioning injectors.
6. Incorrect injector timing.
7. Fuel pump damage.
8. Cylinder valves cracking or clogging.
9. Incorrect valve clearance.
10. Cracked or clogged valves in cylinder head.
11. Excessive carbon buildup in the combustion chamber.
12. Cold operating temperatures.

BLUE EXHAUST SMOKE

Blue exhaust smoke is an indicator that oil is being burned. It can often be observed in a cold engine until it comes up to temperature and the rings expand and reseat back to normal. No quantity of blue smoke is considered normal. Always allow the engine to warm through before applying load:

1. Piston ring wear and blowby.
2. Cylinder wear and glazing.
3. Worn valve guides, stems, or seals.
4. Engine overfilled with oil.
5. Incorrect oil grade for your climate.
6. Oil fuel dilution: injector stuck open.
7. Lift pump worn or damaged.
8. Cylinder glaze burning.
9. Injector ring damaged.
10. Injector pump damaged.
11. Turbocharger seal leaks.
12. Cracked cylinder head.
13. Head oil drain line obstructed.
14. Cylinder head gasket failure: breather clogged.
15. Crankcase over pressure.

7

Engine Operation and Maintenance

There are important points to consider when operating marine diesel engines, and they are also applicable to diesel generators. If you want long-term reliability and maximum life, operating a diesel properly will help:

1. **Warming Up.** Operate at 1,200rpm or less for 5 to 10 minutes; extend this to 15 minutes in very low temperatures. Running a very cold diesel engine stresses everything, from the crankshaft and rods to the camshaft. Proper lubrication and combustion require a warmed-up and thermally stable engine. Slowly bring the engine up to speed and allow it to come to normal operating temperature. This saves fuel and wear. Run your engine at optimum cruising speed. This varies among marine engines, but around 200rpm to 500rpm below peak speed.

2. **Preheating.** If the diesel engine combustion chamber is not heated correctly when starting in cold conditions, cold fuel on the semi-heated glow plugs can result in diesel fuel gelling. This can stick to various internal surfaces and result in damage. When you activate the glow plugs, wait until they heat properly. If you suspect gelling once warmed up, replace your oil filter; it may be partially plugged.

3. **Battery Power.** Battery capacity availability is severely reduced at low temperatures. Make sure the battery is warm, if practicable, and fully charged.

4. **Lubricating Oil.** A diesel engine is up to two to three times harder to turn over at freezing temperatures. Cold and heavy-viscosity oil creates resistance and requires more power to turn over. At the end of every season, preempt cold-start problems with an oil change with the correct temperature-rated oil along with filters. An engine run to ensure that all internal parts are coated is recommended before shutdown. When starting the engine at any time, stop the engine if the oil pressure does not rise to nominal pressure within 10 seconds.

5. **Engine Idling.** Avoid excess idle periods. Prolonged idling causes the coolant temperature to fall below the normal range. This causes crankcase oil dilution due to incomplete fuel combustion; permits formation of gummy deposits on valves, pistons, and piston rings; and promotes rapid accumulation of engine sludge and unburned fuel in the exhaust system. Do not operate at idle for longer than the normal warm-up period, if practicable.

6. **Cooling Water.** If the cooling water is not coming out of the exhaust, or out of the overboard discharge, stop the engine and investigate immediately. Don't allow the pump to run dry or the engine to overheat.

7. **Engine Vibration.** Although an engine vibrates while in operation, it is not the only source of vibration, and you should consider other sources if you are experiencing this. Cavitation is one possible cause, along with worn stern bearings. An incorrect propeller pitch or an unbalanced propeller can cause vibration. You should investigate bent propeller shafts and misalignment of propeller shaft couplings. In some cases, the propeller clearance relative to the hull and stern post can create vibration. A rotating propeller can result in impulses as the blades pass the stern gear components. The rudder, stern post, or supports can also create turbulence if improperly faired, which can create vibration. It is a process of elimination to identify the source.

8. **Shutting Down.** Do not just shut your engine down. Let it cool down at idle speed for several minutes before shutting down. If you have a turbocharger, this allows the bearing temperature to reduce. Also, if you have a turbocharger, do not give the throttle a surge before shutting down. This spools up the turbocharger; when you shut down, the bearings become oil starved, ruining the bearings.

ABOUT ENGINE MAINTENANCE

Maintenance should always be carried out, as a minimum, in accordance with the engine manufacturer's recommendations. I have been extensively involved in preparing and implementing planned maintenance programs on commercial and offshore vessels. Good maintenance will significantly reduce machinery downtime and overall operating costs. A good maintenance schedule, properly recorded, may have significant positive benefits on resale value of the boat.

Note: The following recommended maintenance checklists are also reproduced in the appendix for the convenience of photocopying for use with your own vessel.

ENGINE LAY-UP MAINTENANCE TASKS

___ 1. Run the engine until warm, then drain and replace the oil. Change the filter. Drain and replace the transmission oil when warm. To avoid condensation, top up the oil.

___ 2. Top up the fuel tank, add a stabilizer to the fuel, then circulate the treated fuel and check for leaks. Drain any water within the filter/ separator bowl.

___ 3. Clean the seawater strainer. Rinse the seawater system with freshwater to dilute salt, if possible.

___ 4. Check and renew the coolant additive and run the engine to circulate properly.

___ 5. Lubricate the throttle and gear change linkages using light machine oil.

___ 6. Tape over the exhaust outlet using duct tape.

ENGINE SPRING MAINTENANCE TASKS

___ 1. Change the oil before the start of the season or at the end of the lay-up period, including the oil filters. Run the engine, put in gear, and add some load against the mooring lines. Run up to normal operating temperature.

___ 2. Operate the transmission and throttle cables over the full range several times and lubricate if required.

___ 3. Check and clean the seawater strainer.

___ 4. Check the engine mounts; clean and tighten.

___ 5. Check the electrical connections: clean and tighten.

DAILY AND WEEKLY INTERVAL TASKS

___ 1. Check the lube oil level and top up as required.

___ 2. Check the fuel oil pre-filter for water and drain.

___ 3. Check coolant water level and top up as required.

___ 4. Check transmission oil level and top up as required.

___ 5. Check engine for fuel, oil, and water leaks.

___ 6. Check and top up battery electrolyte levels.

150-HOUR/1-MONTH INTERVAL TASKS

___ 1. Check the fuel filter and replace as required.

___ 2. Check the air filter.

___ 3. Check and clean the crankcase breathers.

___ 4. Perform lubricating oil analysis.

___ 5. Check water system anodes if installed.

___ 6. Check and drain water tanks.

___ 7. Check and clean seawater strainers.

___ 8. Open and close all seacocks and valves five times to exercise them.

250-HOUR/3-MONTH INTERVAL TASKS

___ 1. Change the engine oil.

___ 2. Change the oil filter.

___ 3. Check the condition of the seawater pump impeller.

___ 4. Check and change transmission oil.

500-HOUR/12-MONTH INTERVAL TASKS

___ 1. Check and clean the crankcase vent tube.

___ 2. Inspect all air intake hoses and hose connections.

___ 3. Open and clean water coolers.

___ 4. Check and re-tension rubber drive belt.

___ 5. Check and tighten alternator mounting bolts.

___ 6. Check the coolant water additives.

___ 7. Check and tighten all coupling bolts.

___ 8. Check engine cable harness for chafe.

___ 9. Check and tighten starter and alternator connections.

___ 10. Check and tighten engine negative cable connections.

___ 11. Check the condition of the coolant filler cap.

1,000-HOUR/24-MONTH INTERVAL TASKS

___ 1. Check the engine crankshaft vibration damper.

___ 2. Perform a pressure test of the cooling system.

___ 3. Check and verify engine mountings are clean and tight.

___ 4. Replace the water pump impeller.

2,000-HOUR INTERVAL TASKS

___ 1. Check and adjust the valve clearances.

___ 2. Flush the cooling system and add new coolant.

___ 3. Check all injectors and fuel pumps and overhaul.

8

Diesel Engine Troubleshooting

Injectors are often a prime cause of problems when the engine has high operating hours and an increase in smoke has been noticed or the engine has been running rough. Injectors typically have a reduction in opening pressure, usually 200psi to 300psi; where this is exceeded, injectors may require resetting. Injectors may have poor atomization or be blocked if they have not been serviced for a long period. Air in the fuel system may be a problem, especially in engines that have not been started for long periods. Check that fuel filters are not clogged or that separators are not full of water. Engine timing and low compression problems can be a problem on engines with high operating hours. In some cases, poor fuel quality may be the only cause.

Engine cranks over but will not start:

1. You are using an incorrect starting procedure.
2. The fuel supply valve is closed.
3. The fuel filter is clogged.
4. The fuel lift pump may have a fault.
5. Exhaust line is restricted.
6. Fuel filter is plugged or full of water.
7. Injection pump is not getting fuel.
8. There is air in the fuel line and requires bleeding.
9. There is a faulty injection pump or nozzles.
10. Preheater is not functioning.
11. Starter speed is too low due to low battery charge level.

The engine has a low cranking speed:

1. There is air in the fuel line and requires bleeding.
2. There is an exhaust restriction.
3. The fuel pressure is low with a possible fuel pump fault.
4. There is an injector fault.
5. Starter speed is too low due to low battery charge level.

The engine is hard to start:

1. The gearbox is engaged.
2. There is air in the fuel line and requires bleeding.
3. The starter motor speed is low, battery low voltage.
4. There is water or dirt in the fuel system.
5. The air filter element is clogged.
6. The injection nozzles are clogged or dirty.

The engine starts and then stops:

1. The fuel filter element is clogged.
2. There is air in the fuel line and requires bleeding.
3. The injection nozzles are clogged or dirty.
4. There is an engine timing problem.

The engine has a lack of power:

1. The air intake has a restriction or clogged element.
2. The engine fuel filter or pre-filter is clogged.
3. The engine is overheating.
4. The engine temperature is too low; check the thermostat.
5. The valve clearances are out of specification.
6. There are dirty or faulty injection nozzles.
7. The injection pump timing is incorrect.
8. There is a turbocharger fault.
9. There is a leak in the exhaust manifold gasket.
10. There is a fuel line restriction.

The engine has a loss of compression:

1. There is a worn piston.
2. There are worn or broken piston rings.
3. There is excess cylinder wear.
4. There are leaking valves.
5. There is a leaking or blown head gasket.

The engine has a high temperature:

1. There is no cooling water; faulty water pump impeller.
2. The water pump rubber V-belt is loose or broken.
3. The coolant level is low, with possible hose leaks.
4. The seawater strainer is clogged with debris.
5. The lube oil level is low.
6. The air cleaner filter element is clogged.
7. One of the injector pumps is faulty.
8. The thermostat is faulty.
9. The heat exchanger tubes are clogged.
10. The coolant water cap is faulty.

The engine has a low temperature:

1. The thermostat is faulty.
2. There is high oil pressure.
3. There is a restriction in the relief valve oil passage.
4. The relief valve is out of adjustment.

The engine has a low oil pressure:

1. The lube oil level is low.
2. The lube oil filter element is clogged.
3. The oil pump is faulty.
4. The oil pump suction filter screen is clogged.
5. The oil relief valve is malfunctioning.
6. There is an air leak in the oil pump suction line.
7. The oil pump gears are very worn or damaged.
8. The oil pump cover is loose.
9. The oil pump gaskets are leaking.
10. The oil pressure gauge is faulty.

The engine is using too much oil:

1. There are system oil leaks at joints.
2. The crankcase vent may be dirty or clogged.
3. You are using the incorrect oil viscosity.
4. The air cleaner may be clogged.

The exhaust smoke color is black or gray:

1. The engine has low compression.
2. There is an injector pump fault.
3. The injector nozzles are dirty or faulty.
4. The exhaust line has a restriction.
5. There is a leak in the head gasket.
6. There is an engine timing problem.
7. There is a turbocharger problem or leaking seal.
8. The fuel quality is substandard.

The exhaust smoke color is white:

1. The engine is cold and has low temperature.
2. A valve is stuck.
3. The engine has low compression.
4. There is a leak in the head gasket.
5. The thermostat may be malfunctioning.
6. The injector nozzles may be clogged.
7. There is an engine timing problem.

The engine is knocking:

1. The engine is starting to overheat.
2. There are fuel supply problems.
3. The engine coolant level is low.
4. The engine thermostat is faulty.
5. The fuel filter element is clogged.
6. There is water, dirt, or air in the fuel system.
7. The injector nozzles are dirty or faulty.
8. The engine oil level is low.
9. The fuel injection pump timing is out.

There is high fuel consumption:

1. The air filter is clogged.
2. The engine is overloaded.
3. The valve clearances are incorrect.
4. The injection nozzles are dirty.
5. There is an engine timing fault.
6. There is a faulty turbocharger.
7. The engine temperature is low, thermostat faulty.

The engine is misfiring:

1. The turbocharger is defective.
2. The engine temperature is too low.
3. The engine oil level is low.
4. An injector is either clogged or faulty.
5. There is an injection pump timing problem.
6. The cooling water temperature is low.

9

Engine Starting and Control Systems

The systems that make up a typical diesel engine electrical system include the battery, the engine control panel, the wiring loom, the preheating system, the starter motor and solenoid, the shut-down solenoids, the instrument sensors and transducers, and the alternator. There is a basic sequence of electrical functions that take place when starting the engine. When the key switch is turned to <ON>, this closes the circuit to supply voltage to the control circuit and generally initiates alarms. If no audible or visual alarms activate, it indicates that no power is on. When the key switch is turned to the <PREHEAT> position, this manually or automatically energizes the heating glow plugs or heating elements. When the key switch is turned to <START> or the engine <START> button is pressed, voltage is applied to the starter motor solenoid coil. The solenoid then pulls in to supply main starting circuit current through a set of contacts. When closed, the contacts supply current to the starter motor positive terminal. This then turns the starter motor to start the engine.

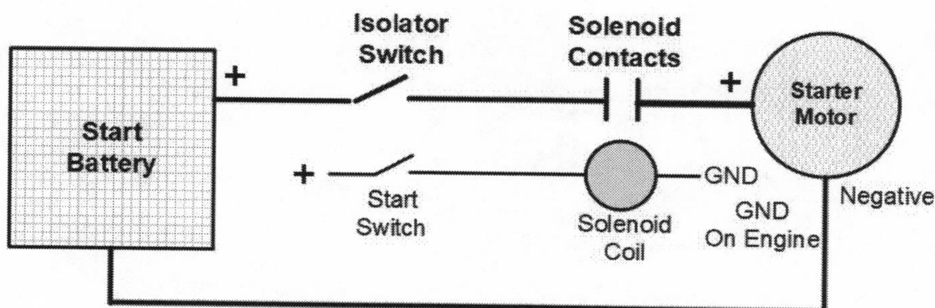

Figure 9-1. Basic Engine Starting System

ABOUT THE ELECTRIC ENGINE STARTER MOTOR

The electric starter consists of a DC motor, a solenoid, and a pinion-engaging drive. The DC motor is typically series wound, as it provides the high initial torque required to exceed friction and inertia (such as oil viscosity) and cylinder compression, and accelerate the engine to a point where self-ignition temperatures and combustion starts—typically in the range of 60rpm to 200rpm, depending on whether glow plugs are used. The starter motor torque is transmitted by the pinion and ring gear on the flywheel. The drive gear pinion has a reduction gear of around 15:1.

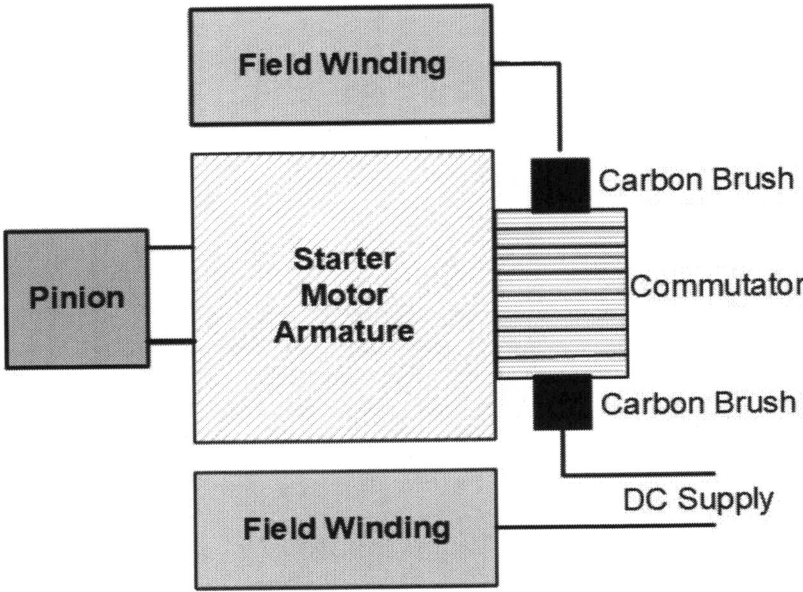

Figure 9-2. Starter Motor Diagram

ABOUT STARTER MOTOR SOLENOIDS

The solenoid is a large, high-current relay that consists of a coil and armature, along with moving and fixed contacts. The solenoid is mounted directly to the starting motor housing, which reduces cables and interconnections to a minimum. When the solenoid coil is energized by the starting circuit, the solenoid plunger is drawn into the energized core; this closes the main contacts to supply current to the starter motor. On some starters the solenoid has a mechanical function. The solenoid activates a shift or engaging lever to slide the overrunning clutch along the shaft to mesh the pinion gear with the flywheel; when engaged, the starter motor then turns the engine, so meshing occurs before starting.

STARTER MOTOR TYPES

The motor consists of four pole shoes or magnets; some use permanent magnets. The poles are fitted with an excitation winding, which creates the magnetic field when current is applied. The rotating part, called the armature, incorporates the commutator. The four carbon brushes provide the positive and negative power supply. There are four basic DC motor types in use, and they are based on connection of the field windings. The field windings are connected either in series or parallel with the armature windings.

1. **Shunt (Parallel)-Wound Motors.** The motor operates at a constant speed irrespective of loads applied to it. It is the most common motor used in industrial applications, and is suited to applications where starting torque conditions are not excessive.
2. **Permanent Magnet–Excited Motors.** The permanent magnet starter offers the advantages of reduced weight, physical size, and less heat generation than normal field type starters. Current is supplied via the brushes and commutator directly to the armature. Another feature is a reduction gear, which allows faster speeds and increased torque.
3. **Series-Wound Motors.** This type of DC motor has a speed characteristic where the speed varies according to the load applied, meaning speed increases with load decrease.
4. **Compound Motors (Series/Shunt-Wound).** This configuration is often used on large starter motors. It combines the advantages of both shunt and series motors, and is used where high starting torques and constant speeds are required.

ABOUT PINION-ENGAGING DRIVES

The pinion-engaging drive is located within the end shield assembly of the starter and consists of the pinion-engaging drive and pinion, the overrunning clutch, and the engagement lever or linkage and spring. When the motor operates, the drive gear meshes with the ring gear or flywheel teeth to turn the engine and then disengages after starting. The overrunning clutch has two important functions. The first is to transmit the power from the motor to the pinion; the second is to stop the starter motor armature from overspeeding and being damaged when the engine starts. Preengaged starters generally use a roller-type clutch, while larger multi-plate types are used in sliding gear starters.

ABOUT STARTER TYPES

There are several types of starters in use, the most common being the overrunning clutch starter; and inertia-engagement Bendix drive is now less common. There are four basic groups of starter motors.

PREENGAGED (DIRECT) DRIVE STARTERS

The most common type of starter motor is the solenoid-operated direct drive unit, and the operating principles are the same for all solenoid-shifted starter motors. When the ignition switch is placed in the <START> position, the control circuit energizes the pull-in and hold-in windings of the solenoid. The solenoid plunger moves and pivots the shift lever. This moves the pinion along the shaft to mesh or engage with the flywheel toothed ring gear. When the solenoid plunger is moved all the way, the contact disc closes the circuit from the battery to the starter motor. Current now flows through the field coils and the armature. This develops the magnetic fields that cause the armature to rotate and turn the engine.

GEAR REDUCTION STARTERS

Some manufacturers use a gear reduction starter to provide increased torque. The gear reduction starter differs from most other designs in that the armature does not drive the pinion directly. In this design, the armature drives a small gear that is in constant mesh with a larger gear. Depending on the application, the ratio between these two gears is between 2:1 and 3.5:1. The reduction allows a small motor to turn at higher speeds and greater torque with less current draw. The solenoid operation is similar to that of the solenoid-shifted direct drive starter in that the solenoid moves the plunger, which engages the starter drive.

SLIDING GEAR DRIVE STARTERS

These two-stage starters have either mechanical or electrical pinion rotation. The electrical units have a two-stage electrical pinion-engaging drive. The first stage allows meshing of the starter pinion without cranking the engine over. The second stage starts when the pinion fully travels and meshes, and then allows full excitation and current flow to the starter motor. The first stage of mechanical units has a solenoid switch, which pushes forward the pinion-engaging drive via a lever. When pinion meshing occurs, current is applied to the starter via the solenoid switch.

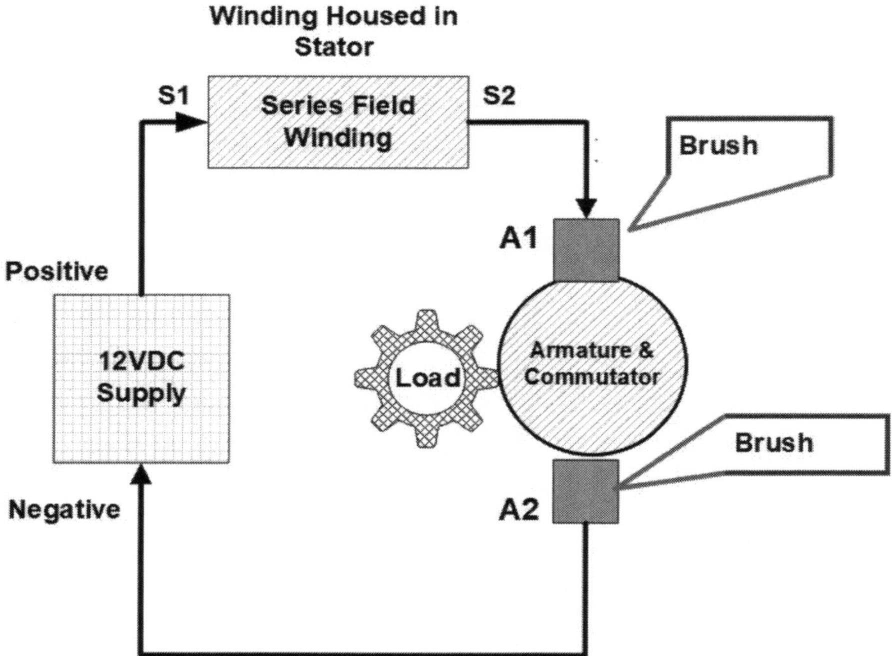

Figure 9-3. DC Series Motor Diagram

BENDIX DRIVE INERTIA STARTER

The Bendix friction-clutch mechanism drive was developed in the early twentieth century. These starters use a drive friction clutch, which has a drive pinion mounted on a spiral-threaded sleeve. The sleeve rotates within the pinion, moving the pinion outward to mesh with the flywheel ring gear; the impact of this meshing action is absorbed by the friction clutch. Once started, the engine turns at a higher speed and drives the Bendix gear at a higher speed than the starter motor. The pinion then rotates in the opposite direction to the spiral shaft and disengages. The drive pinion being thrown out of mesh and then stopping is a common fault. Always wait several seconds before attempting to restart or the drive mechanism may be damaged. Another fault is when the pinion does not engage after the starting motor is energized. If the starter emits a high-pitched whine, turn off the ignition immediately as the unloaded DC starter motor will overspeed and be seriously damaged. Problems can be minimized by ensuring that the sleeve and pinion threads are clean and lubricated so that the pinion engages and disengages freely. The Bendix gear, shaft, bearings, and end plates can be cleaned of dried grease with a suitable cleaner such as CRC Lectra Clean. You can lubricate with WD-40 multipurpose penetrating oil or a fine 3-in-1 sewing machine oil. There are starter motor–specific greases with cold temperature capability. The PolySi G-MAN Lubricant, PST-433 extreme low temperature grease is a well-known product. If you lay up your boat every winter in a very cold climate, consider this or an equivalent product.

STARTER MAINTENANCE AND TROUBLESHOOTING

Starter installation is generally limited to two factors. The first is being mechanically secure, and the second is that the attached electrical supply cables are of the correct current rating and that terminal nuts are properly torqued so they do not work loose and create a high-resistance joint. The negative cable should be attached as close as practicable to the starter motor. Starter motor design is generally robust, as it must withstand the shocks of meshing, engine vibration, salt- and moisture-laden air, water, oil, temperature extremes, high levels of overload, and so on. A common problem, especially on idle vessels, is the buildup of surface corrosion or accumulated dirt on the shaft and pinion gear assembly; lack of lubrication causes seizure or failure to engage. It is good practice to remove the starter every twelve months and clean and lightly oil the components according to the manufacturer's recommendations.

STARTER MOTOR MAINTENANCE

Problems often occur with seized brushes, and this is primarily caused by lack of use. Always manually check that brushes are moving freely in the brush holders and that the commutator is clean. Remove all dust and particles using a vacuum cleaner. If badly soiled, wash out with a quality spray electrical cleaner. Never clean or polish the commutator with any abrasive materials.

STARTER TROUBLESHOOTING

Many are familiar with the silence and loud click when the start solenoid operates but the starter fails to turn over. The main cause is a bad negative connection, which results in lower voltage. Other possible causes include a poor positive connection caused by loose or dirty connections, or a solenoid plunger sticking and not closing fully, preventing the main contacts from closing.

• NEGATIVE CONDUCTOR CONNECTIONS

To prevent the connection becoming loose from vibration, the main negative cable should be secured on the engine using a spring washer. The main engine negative cable connection is prone to vibration from the engine operation. Cables frequently come loose, causing starting problems, with high resistance and intermittent equipment operation, interference, and, in some instances, alternator failures. In most cases, they are simply fastened to a convenient bolt, which is not acceptable if you value reliability. The mating surface must be cleaned to ensure a high-quality, low-resistance electrical contact. Install a spring washer to maintain tension.

ABOUT PREHEATING GLOW PLUGS

Some engines will not start without preheating, and extended engine starting turnover times may overheat and damage the starter. Many marine diesels have glow plug heaters installed. They preheat the air in each cylinder to facilitate starting. In cold weather, this will dramatically decrease the electrical power requirements to start the engine.

1. **Activation.** Prior to engine starting, the plugs are activated for an operator-selected time period or interlocked to a timer, and typically ranges 15 to 20 seconds.

2. **Power Consumption.** The glow plugs can draw relatively large current levels for a short time. If your battery is low, allow a few seconds after preheating before starting; this enables the battery voltage to recover from the heater load.

ABOUT AIR INTAKE HEATERS

These grid-resistor heaters are installed in the main air intake of indirect injection (IDI) engines, and there is normally only one heating element.

ABOUT PREHEATING CONTROL

Many preheating circuits have relays, either timed or untimed. Timed relays are often a common cause of failure. It is advisable to have a straight relay with a separate switch, and simply preheat manually for 10 to 15 seconds and then start.

PREHEATER TROUBLESHOOTING

Preheater glow plug connections must be regularly checked if they are to function properly. The connections must be cleaned and tightened every six months. The insulation around the glow plug connections must be cleaned. It is a common fault to have voltage leakage to ground caused by tracking across oil and sediment to the engine block, with a serious loss of preheating power. The glow plugs should be removed and cleaned yearly. Take care not to damage the heating element.

There is no preheating:

1. Power is lost (fuse failure).
2. There is connection fault on engine to first plug.
3. There is relay failure.
4. Connection on ignition switch is disconnected.
5. Circuit has short circuit to engine block.

There is partial preheating:

1. One or more glow plugs have failed.
2. Glow plug interconnection has failed.
3. Dirt around glow plug is causing surface tracking.

ENGINE STARTING SYSTEM DIAGRAMS

In the previous edition of this book, I included simplified wiring diagrams for a variety of engines. Over the years, there have been so many engine variants and technology changes that this is no longer practical. Always check the diagrams supplied in the operator's manual for your specific engine model; you can download a copy from the manufacturer's website to your computer, tablet, or notebook. Make sure you have the correct circuit diagram for the installed engine; I suggest printing and laminating a copy. Wiring varies considerably, even between older and newer engine models. The following table gives equivalent color codes for various manufacturers. The illustration is a typical Volvo engine starting circuit; although newer engines have different arrangements, it does illustrate the various elements.

Purpose	**US Codes**	**Yanmar**	**Volvo**
Ignition Start	yellow/red	white	red/yellow
Ignition Stop	black/yellow	red/black	purple
Diesel Preheat	orange/yellow	blue	orange
Negatives	black	black	black
Alternator Light	brown	red/black	brown
Tachometer	gray	orange	green
Oil Pressure Gauge	light blue	yellow/black	light blue
Oil Pressure Alarm	white/blue	yellow/white	blue/white
Water Temperature Gauge	tan	white/black	light brown
Water Temperature Alarm	white/brown	white/blue	brown/white

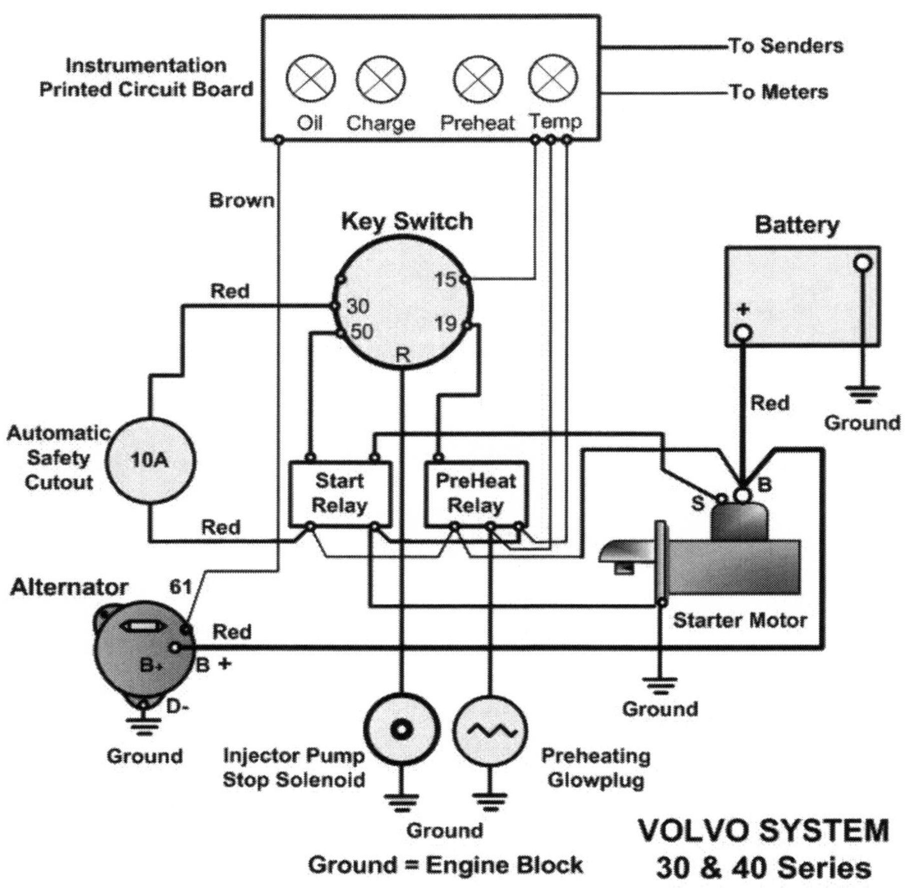

Figure 9-4. Typical Volvo Starting Circuit

BASIC STARTING SYSTEM REQUIREMENTS

The power supply to the engine starting system should have an isolator installed as close as practicable to the battery in both the positive and the negative conductor. The battery isolator should be accessible. Short-circuit protection is not required on a start circuit. The isolator must be rated for the maximum current of the starting circuit. The main starting circuit positive and negative conductors should be rated so as not to exceed a 5% voltage drop at full rated current. The main start cables should have minimal voltage drop at full rated current. Cables should be kept as short and as large as possible to minimize resistance losses and maximize power availability.

ENGINE ELECTRICAL TROUBLESHOOTING

Always check that the engine preheating system is operating and cranking speed is correct, as low battery voltages may not be turning the engine over fast enough. Check that the solenoid is operating at the starter motor; an audible click indicates functionality. The most common faults in diesel engine electrical systems are low battery voltages or bad electrical connections.

The engine will not crank over:

1. Battery isolator switch is not in the ON position.
2. The throttle lever is not in the neutral position.
3. The neutral position interlock switch is sticking.
4. The shaft brake interlock switch is sticking.
5. The control power has failed (plug connection).
6. The control system fuse has blown.
7. The battery voltage is very low.
8. The stop solenoid has seized.
9. There are loose starter motor terminals.
10. The start circuit relay is faulty.
11. The start button is faulty.
12. The key switch is faulty.
13. The starter solenoid connection has broken or is loose.
14. The starter solenoid coil has failed.
15. The stop button is jammed in.
16. The starter motor has seized.

Solenoid operates but starter motor will not turn:

1. The solenoid main contacts are faulty or jammed.
2. The starter motor bearings have seized.
3. The starter motor has a mechanical failure or has seized.
4. The starter motor brushes have jammed.
5. The starter motor windings have failed.

Low engine cranking speed:

1. The battery voltage is low.
2. The battery terminals are not tight.
3. The starter motor is faulty.
4. The starter motor electrical connections are loose.
5. The negative connection on the engine is loose.

Engine will not stop:

1. The stop solenoid connection is loose.
2. The key switch is faulty.
3. The stop solenoid has seized.
4. There is a wiring harness connector fault.
5. There is a control fuse failure.

Starting battery is undercharging:

1. There are too many auxiliary loads connected.
2. The engine is running at low idle speed.
3. The alternator main connection is not tight.
4. The battery is defective and sulfated.
5. The alternator diodes are faulty.
6. The alternator regulator is faulty.
7. The alternator V-belt is loose or slipping.
8. The alternator mounting bolts are loose.

Battery water consumption has increased:

1. The ambient operating temperature has increased.
2. The alternator regulator is faulty with high output voltage.

Starting battery will not charge:

1. There are loose or corroded battery connections.
2. The alternator main positive output cable connection is not tight.
3. The main negative connection is not tight.
4. The alternator mounting bolts are loose.
5. The batteries are heavily sulfated.
6. The alternator V-belt is loose or slipping.
7. The alternator or regulator is faulty.

ENGINE STARTING SYSTEM CONFIGURATIONS

There are several engine starting configurations and arrangements.

REMOTE BATTERY ISOLATORS

Many vessels have simple mechanical isolation switches to isolate the engine starter motor power supply. In many cases, remote isolation using relay-type isolators is used. The control relay may be operated from a separate switch or interlocked to the main key switch so that power is applied when the switch is turned.

DUAL- OR TWO-POLE ENGINE SYSTEMS

In many engines that have dual-pole isolated systems, two battery isolation relays are installed—one on positive and one on negative. The relay coil is connected to the alternator D+ terminal, and this energizes the coil when the alternator is operating. In remote isolation relay systems, one relay can be used to energize both switches.

PARALLEL BATTERY STARTING SYSTEMS

Some vessels have a 12V power system and a 24V engine system. The batteries are configured through a relay so that the parallel-connected batteries are series connected to 24V when the engine start switch is operated.

PARALLEL CONNECTED STARTERS

Larger engines sometimes use two starter motors, which keeps the size down. The system uses a large-capacity double-acting relay to supply current to both starter motors simultaneously.

ABOUT NEUTRAL SWITCHES

This is a switch installed within the mechanical remote engine control unit. They are used to prevent starting of the engine if the gearbox is engaged, which could cause serious damage. A microswitch is used to activate a relay that is inserted in the starting control circuit. If you have to move the throttle a few times to get a start, the switch and adjustment require readjustment.

ENGINE AND TRANSMISSION CONTROL SYSTEMS

While traditional engine controls have centered around push-pull Morse control cables, there is an ever-increasing use of fully electronic systems such as those from ZP-Mathers, Teleflex Morse, Twin Disc, and Kobelt. Systems may have an electronic throttle and shift arrangement or a mechanical shift with electronic throttle. Teleflex Morse has the Teleflex Intelligent System (TIS) MagicBus, which uses the Controller Area Network (CAN) or CAN bus network protocol. In this system all engine control and monitoring signals, data, and transducer inputs and outputs share the same network.

ABOUT THE CONTROL HEAD OR STATION

This comprises the throttle and gear change levers. The control head outputs a variable DC voltage to the actuators. Some systems may use a pulse width modulated (PWM) signal. The DC voltage value corresponds to the position of the control lever. The control heads incorporate a light emitting diode (LED) visual and audible warning indication and the control transfer button. The control head has three separate circuits and one common circuit: the potentiometer, transfer button, sound transducer, and the LED indicator circuits.

ABOUT ACTUATORS

Actuators have an integral control circuit board and accept the variable DC command voltages, converting them into mechanical outputs. These may be solenoid valves and linear actuators, with drive motors and coils, which operate the clutch or transmission ahead or astern, and for speed control to the fuel rack. The feedback loop to the control head or control modules is via gear-operated potentiometers. The control systems are essentially closed-loop controllers. A set point is used; for example, the throttle is moved to the required position, and the control system, via actuators, increases engine speed to match. The engine speed feedback signal is compared to the set point; when there is no differential, no further speed signals are sent to the actuators. Synchronization for actuators and processors is required in twin-engine installations.

SYSTEMS WIRING

Wiring interconnects the control heads, actuators, and any electronic control module (ECM). These include the power supply cable, start interlock cables, and multicore control cables. The start interlock cables are connected to the starter solenoid and the actuator via a normally open relay. The multicore cable connects the control head to the actuator.

CONTROL MODULES

Some systems may have a separate control module. This comprises a power supply unit (PSU) and the microprocessor central processor units (CPU) that input, process, and output signals to the actuators. They perform the input signal conversion from the throttle and gear changes, and output signals to the actuators via a distributed control network. Modules have displays to enable performance monitoring or troubleshooting using fault codes.

SAFETY AND CONTROL SYSTEMS

Most mechanical and electrical controls have several safety systems and operating features incorporated. Redundancy is important within any propulsion control system. Manufacturers of control systems and engine management systems incorporate backup sensors in case the primary ones fail. This can include speed and injection timing, throttle position, boost pressure for fuel air ratio control, coolant temperatures, and lubricating oil pressures. Systems may have combination LED fault code and audible warning codes to indicate status and faults.

1. **Neutral Interlock.** A microswitch activates a start-blocking relay that prevents the starting of the engine when the engine transmission is engaged, which is selected ahead or astern. This will prevent engine start until the clutch is disengaged. The control system has to be switched on and command acceptance carried out. Switches and relays can cause problems, so always look at this first if you cannot get the engine started.
2. **Shaft Brake Interlock and Control.** Shaft brakes incorporate a sensor that interlocks engine starting until the brake is released. Switches can cause problems and should be checked.
3. **Reduction Gear Oil Pressure Interlock.** If gear oil pressure fails to build up or falls, the engine speed is reduced to idle. It is important to ensure oil filters are clean and oil levels are correct to avoid this occurring at speed.
4. **Drive Train Reversal.** This is called crash reversal, and the control sequences the shift and speed functions so that when an emergency reversal is requested, it provides the shortest possible time without damage to the drive train or stalling of the engine.
5. **Warm-up Mode.** This enables engine speed adjustment with the engine in the neutral (Neutral Fast Idle Mode) control position.

ABOUT ENGINE SYNCHRONIZATION

This is used to decrease vibration and noise and reduce fuel consumption, and is an automatic function. In most cases, a leading engine is nominated, and the following or slave engine speed is adjusted to match it. Some units use engine tachometers, while others use a proximity sensor on the shafts that increases synchronization accuracy.

TROLLING OR SLOW SPEED MODE

This is for speed control that is lower than normal engine idle speeds. A trolling valve comprises pressure-reducing solenoid valves. The valves are operated by a servo, which operates via a control signal. When actuated, a solenoid valve opens, allowing fluid to go to the pressure reduction valve.

INSTALLATION NOTES

It is recommended that two power supplies be provided via a changeover switch so that control is available if a loss of one battery bank occurs. Typical power consumption is 10A, and a stable power supply is required. This should not be taken from the engine starting batteries. In noisy electrical environments, a separate and isolated battery supply should be considered. Shielded, twisted-pair data cables are used to prevent interference. The CPU should be isolated from any vessel grounding system, in particular steel or alloy boats. Shielding at actuators and control stations should not be grounded to prevent circulating ground currents. Grounding of the shield is at the CPU only. Excess cable length should not be coiled up, but cut to fit.

ABOUT ENGINE SYNCHRONIZERS

While some integrated engine control systems incorporate the synchronization function, discrete systems for retrofitting are available. In these systems the actuator is placed in the helm throttle control to the engine control cables. Speed signals are picked up from the nominated lead and follow engines. When synchronization is required and activated, the system will match engine speeds whenever the throttles are within 15% of each other. Correct installation is essential, and electrical connections must be tight, in particular the ground. Power consumption is low, at just under 1A when in control mode.

ABOUT ELECTRONIC CONTROL MODULES

This is a computer with operating software designed for a specific engine, which performs all the monitoring and control functions. The ECM supplies power to the electronics, processes sensor input information, outputs actuator signals, processes and outputs monitoring information, performs diagnostic routines, and acts as a governor to control engine speed. When the ECM receives a requested throttle speed signal, it controls fuel injection and maintains engine speed by comparing required speed with actual speed that is fed back from the speed sensors. Speed is controlled through injection timing and fuel quantity.

ABOUT ENGINE SPEED MONITORING

The speed-timing sensor consists of a permanent magnet and a coil. The teeth on the camshaft pass through the sensor magnetic field to generate a voltage, and the time between voltage pulses are counted to derive speed. These are a two-wire device that does not require a power supply. There are typically two sensors: one on the camshaft and one on the crankshaft. The camshaft unit is used for injector timing, and the crankshaft one is used for more accurate speed measurement. In two-sensor systems, one acts as a backup; in single-sensor systems on older engines, failure causes the engine to shut down. These sensors are critical to the primary speed and timing function of the ECM, which governs engine operation.

THROTTLE POSITION SENSING

This is normally the throttle lever, and it sends the requested speed signal to the ECM. It outputs a PWM signal to the ECM. It has a three-wire input—the supply voltage of 8V, the ground reference, and the output signal. The PWM output is a constant frequency square wave, either full voltage or zero, on or off signal. The duty cycle is the percentage of ON time. In idle mode this is 10% to 22%; in high idle it is 75% to 90%. The ECM monitors the duty cycle, and if less than 5% or greater than 95%, it indicates a fault.

INSTALLATION AND TROUBLESHOOTING

The reliability of systems depends on proper installation, location of equipment away from moisture and heat, and a clean electrical power supply. Units must withstand higher machinery space temperatures, vibration, electric currents, radio frequency interference (RFI), and electrostatic discharges. Actuators are normally bonded to maintain equipotential levels. Many cables from control units and engine management systems use multicore cables; most are screened and screens must be terminated correctly. It is important to identify locations where cables may chafe, have tight bends, or where mechanical damage is possible. Relay junction boxes must be mounted in locations that minimize mechanical damage, vibration, and heat exposure. It is common to blame injectors for engine conditions such as misfiring, low power, and rough or erratic operation when instead the cause can be something simple like fuel filter blockages. In many cases simple faults such as deteriorated solenoid connections or wiring faults or loose connectors are a cause for starting problems, so check them first, clean, and retighten.

10

Engine Instrumentation Systems

Engine instrumentation is crucial to ensuring that engines operate correctly within the designed parameters. Instruments may consist of a bank of discrete analog meters or an integrated system with digital and visual screen displays, and most manufacturers have such systems. The latter is becoming more prevalent and consists of trend analysis, alarm set-point management, alarm logging, and other advanced features. Check all sender-unit terminals and connections regularly along with a test of all alarm functions, preferably before you start your voyage. Typical of the new-generation digital screens is the Yanmar multifunction display. The processor is able to display virtual gauges as well as alarms and diagnostic troubleshooting codes. Parameters include engine speed, engine coolant temperature, engine hours, oil pressure, and engine load, along with wind, speed, depth, and Automatic Identification System (AIS) data. The ultrawide full-color display has a 170-degree viewing angle. The NMEA 2000 connectivity allows data transfer to all compliant devices; an example is the Raymarine i70 control heads, which can display engine operating data. The keys on many control stations are made from waterproof silicon.

ABOUT TEMPERATURE MONITORING

The main temperature monitoring points utilize similar thermistor pellet or resistance temperature detector (RTD) sensor types. These are a resistor that has a change in resistance value when the temperature changes. They are used on lubricating oil, transmission oil, seawater and freshwater coolants, fuel temperature, after-cooler and turbocharger inlet air, and so on.

WATER TEMPERATURE ALARMS

These alarm devices consist of a bimetallic element that closes when the factory-set temperature is reached. In many sensors they are incorporated into the same sensor unit. To test, simply remove the connection from the sender terminal and touch it on the engine block to activate the alarm. The sensor has two terminals: "G" is used for the meter and "W" is used for the alarm contact. In many boats, damage often occurs because the alarm did not function or was not noticed. The first reaction is often "What's wrong with the alarm?" not "What's wrong with the engine?" It is good practice to add a very loud audible alarm, as some of the engine panel units are difficult to hear at times over the ambient engine noise.

ABOUT OIL PRESSURE ALARMS

The monitoring of oil pressure is fundamental to proper operation of any diesel engine. This includes the lubricating oil and filter differential pressures, diesel fuel and filter differential pressures, seawater and freshwater coolants, turbocharger charging air pressure air inlet pressures, gearbox and transmission oil pressures, and engine crankcase pressures. A pressure alarm either is incorporated into a gauge sender unit or is installed as a separate device. It consists of a pressure-sensitive mechanism that activates a contact when the factory-set pressure is reached. They are grounded to the engine block on one side, and operation grounds the circuit, setting off the panel alarm.

ABOUT OIL PRESSURE MONITORING

The oil pressure sender unit is a variable resistance that alters proportional to pressure. Oil pressure sender units should be removed every year and any oil sludge cleaned out of the fitting, as this can commonly clog, causing an inaccurate or no reading. Low oil pressure readings are caused by low lube oil level or a clogged oil filter causing a lowering in oil pressure. A faulty oil pump can cause a lowering in pressure or a rise in oil temperature caused by an increase in engine temperature or an oil cooler problem. Sender units are often poorly grounded or Teflon tape is improperly applied to threads to make a high-resistance contact.

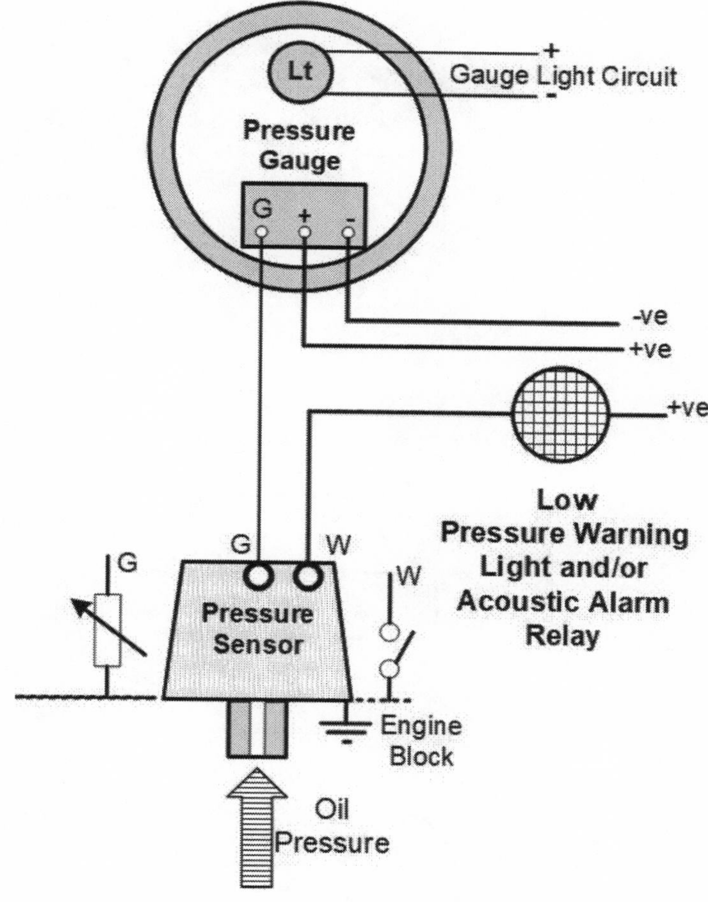

Figure 10-1. Oil Pressure Monitoring

WATER AND OIL TEMPERATURE GAUGES

The main temperature monitoring points utilizing the same sensor types include lubricating oil, transmission oil, seawater and freshwater coolants, fuel temperature, after-cooler, and turbocharger inlet air. Monitoring water temperature is essential to the safe operation of the engine; temperature extremes can cause serious engine damage or failure. Sender units are resistive and output a resistance proportional to temperature in a nonlinear curve. If the gauge readings are not correct and a gauge test shows it to be good, check the sensor. Before you check the sender unit, the main causes of high temperatures are the loss of freshwater cooling caused by a faulty water pump impeller, a loose rubber drive belt, low water levels, fouled coolers, and increases in combustion temperatures. Loss of saltwater cooling is caused by a blocked intake or strainer faulty water pump impeller, clogged cooler, or aeration caused by a leak in the suction side of the pump. Increased engine loadings caused by adverse tidal and current flows or overloading also lead to this condition. Sender units are often poorly grounded or Teflon tape is improperly applied to threads to make a high-resistance contact.

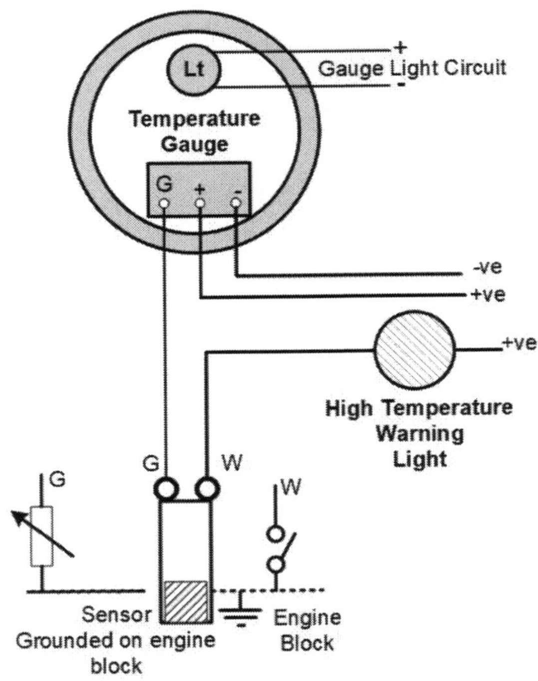

Figure 10-2. Water Temperature Monitoring

EXHAUST GAS TEMPERATURE MONITORING

Exhaust gas temperature monitoring is used in commercial ships and is recommended on all motor-sailers, motorboats, trawler yachts, and large sailing yachts with high output engines. Engine problems are easier and faster to identify than water temperature and oil pressure monitoring. These can be problems within the cooling water system; increased engine loads that may be caused by adverse tidal and current flows; air intake obstructions caused by clogged air filters or, where installed, blocked air coolers; and combustion chamber problems caused by defective injectors, valves, and so on. Larger-engine vessels will often have cylinder monitoring; this allows identification of problems specific to cylinders to be identified and monitored. Smaller engines may have a sensor installed on the main exhaust manifold. Pyrometer compensating leads and wiring should be routed clear of other cables to avoid induction and inaccurate readings.

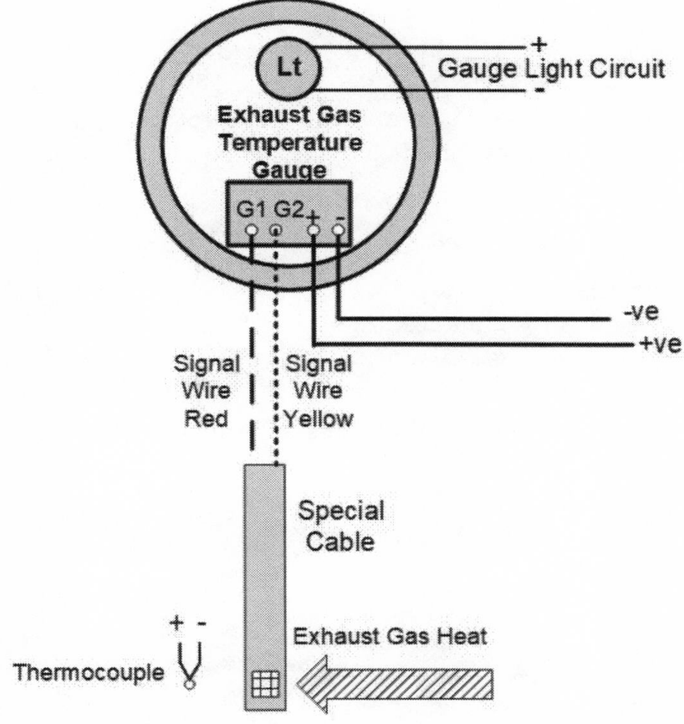

Figure 10-3. Exhaust Gas Temperature Monitoring

EXHAUST GAS TEMPERATURE SENSOR OPERATION

Exhaust temperature sensors are called thermocouples, or pyrometers. These sensors consist of two dissimilar metals—iron/constantine, copper/nickel, platinum/rhodium, nicrosil/nisil, nickel/aluminum—which at the junction will generate a small voltage proportional to the heat applied to the sensor. The voltage is measured in millivolts (mV). The typical thermocouple consists of a sensing junction and a reference junction. The open circuit voltage is measured with a high-impedance voltmeter and is the temperature difference between the sensing junction and the reference junction. The thermocouple junction is often called the "hot junction." The compensating cables between the junction and the measurement meter are electrically matched to maintain accuracy; they are polarity sensitive, so must be connected positive to positive.

ENGINE TACHOMETERS EXPLAINED

The tachometer is used to monitor engine speed, differential or synchronization, shaft revolutions, and turbocharger speed. This information enables decisions to be made on fuel consumption and vessel performance. There are a several tachometer types, based on the type of sensing system.

ABOUT THE GENERATOR TACHOMETER

These tachometers take a signal from a mechanically driven generator unit. The generator outputs an alternating current (AC) voltage proportional in amplitude to the speed, and this is decoded by the tachometer. Variations in speed give a proportional change in output voltage and a change in meter reading. The most common fault on these units is damage to the drive shaft mechanism. These are typically marked "G" and negative (-).

ABOUT THE INDUCTIVE TACHOMETER

These tachometers have an inductive magnetic sensor that detects changes in magnetic flux as the teeth on a flywheel move past. This sends a series of on/off pulses to the meter, where they are counted and displayed on the tachometer. Ensure that the sender unit is properly fastened. A common cause of failure is damage to the sensor head if it strikes the flywheel when adjusted too close or becomes loose. These are typically marked "W" and "G."

ABOUT THE ALTERNATOR TACHOMETER

These tachometers derive a pulse from the alternator AC winding, typically marked "W." The alternator output signal is a frequency directly proportional to the engine speed. The pickup is taken from the winding star point or one of the unrectified phases. If the alternator is faulty, there is no reading. There are a number of different alternator terminal designations used by various alternator manufacturers; the main ones are W, STA, AC, STY, and SINUS. If there is no output terminal, a connection will have to be made.

SYNCHRONIZATION TACHOMETERS

The synchronizer, or differential tachometer, is used to show the precise speed difference between each engine in a twin-engine installation. Use of the meter allows precise balancing to be carried out.

BILGE AND TANK LEVEL MONITORING

The monitoring of onboard fuel, water, and bilge levels is an essential task. The majority of tank sensors currently in use operate by varying a resistance proportional to tank or bilge level. The two basic sensor types are described below.

ABOUT THE IMMERSION PIPE TYPE

This sensor type consists of a damping tube with an internal float that moves up and down along two wires. These units are only suitable for fuel tanks. The big advantage with these sensor types is that they are well damped; therefore, fluctuating readings are virtually eliminated.

LEVER-TYPE SENSORS

The lever-type system consists of a sensor head located on the end of an adjustable leg. The sensor head comprises a variable resistance and float arm pivot. As the float and arm move relative to fluid levels, the resistance alters and the meter reading has a corresponding change. Typical resistance readings are in the range 10Ω to 180Ω. Lever-type units should be installed longitudinally, as athwartships orientation can cause problems with vessel rolling. Where these units are used for fuel or water, the primary difference is that for water sensor units, the variable resistance is located outside the tank to avoid water ingress problems, while the fuel unit has a resistance unit in the tank.

ABOUT CAPACITIVE SENSORS

This type of transducer operates on the principle that the value of a capacitor is dependent on the dielectric between plates. The sender unit measures the capacitance difference between air and the liquid. The sensing circuit outputs a voltage proportional to the level, in the typical range of 0V to 5V. The most common fault in these systems is water damage to the circuit board, usually because of tank condensation.

ABOUT PRESSURE SENSORS

These sensor types are very accurate and less prone to damage. The transducers are either placed at the bottom of the tank or on a pipe that is located at the tank bottom. The sensors output either a 4mA to 20mA or 0.6V to 2.6V proportional to the pressure of the fluid in the tank. The pressure value is proportional to the tank volume. If the sensor is located on a small pipe, it may become clogged.

ABOUT VOLTMETERS

Many instrument panels incorporate a voltmeter to indicate the state of charge of the battery or the charging voltage. As they have a coarse scale, they are only partially useful in precisely assessing battery voltage states, but they are a useful indicator on the charging system. Many voltmeters have a colored scale to enable rapid recognition of condition—red for under- or overcharge and green for proper range.

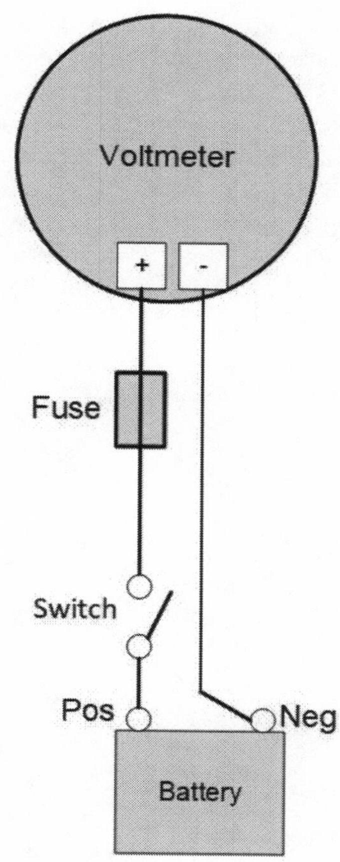

Figure 10-4. Voltage Monitoring

ABOUT CHARGING AMMETERS

Charging ammeters are an indication to the level of charge current from the alternator. In most cases the level of current is not as important as observing current flowing. There are two types of ammeters: the inline, or series, ammeter and the shunt ammeter.

SERIES AMMETERS

This ammeter type is connected directly in series with the load. It has the main charge alternator output cable running through it. In many cases, the long run to a meter causes unacceptable charging system voltage drops and undercharging. A problem with installing such ammeters on switch panels is that the charge cables are invariably run with other cables, often causing radio interference. If you are going to install this type of ammeter, ensure that the meter is mounted as close as possible to the alternator. If these ammeters start fluctuating at maximum alternator and rated outputs, this is generally due to voltage drops within the meter and cable. The underrating of connectors is a major cause of problems.

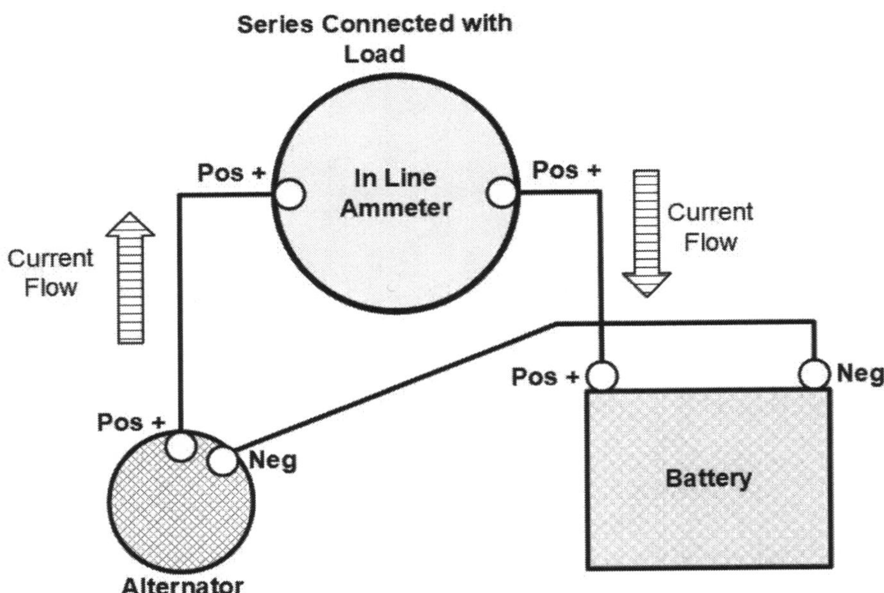

Figure 10-5. Current Monitoring

SHUNT AMMETER OPERATION

The shunt ammeter overcomes the problem of long cable runs and voltage drops associated with a series ammeter. The ammeter shunt is essentially a resistance inserted in the line. Sense cables are connected across it and can be run to the meter location without voltage drop problems, as the output is in millivolts. The ammeter shunt must always be rated for the maximum alternator output. In many installations, this is not done, and the shunt is often damaged, with significant voltage drops in the charging line.

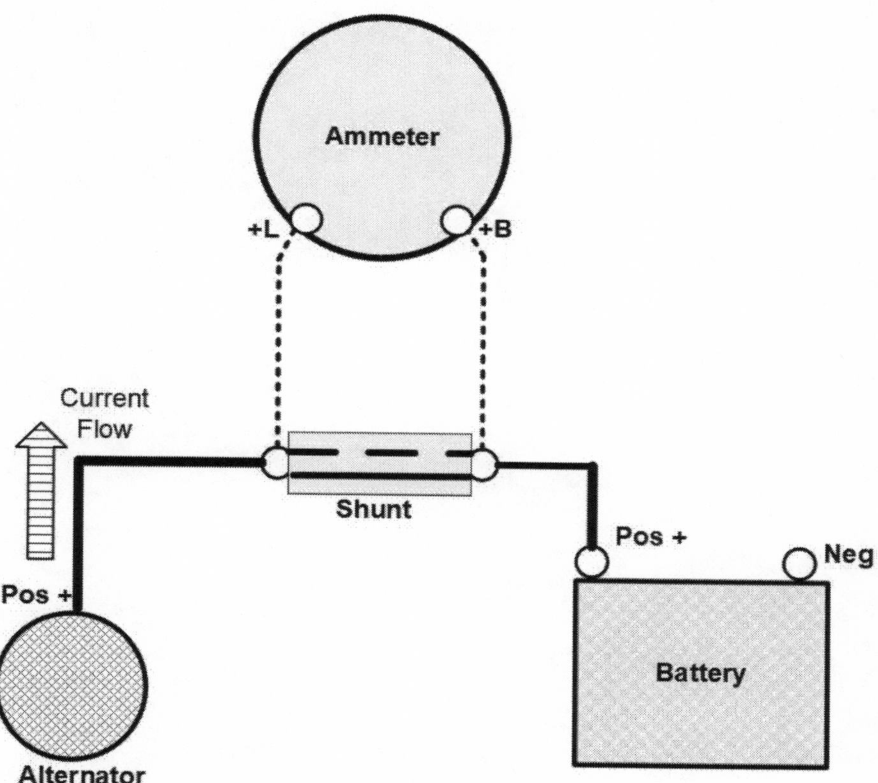

Figure 10-6. Shunt Ammeter Diagram

ABOUT HOUR COUNTERS AND CLOCKS

Operating hour counters are essential to keeping a record of maintenance intervals. Essentially, it is a clock that is activated only when the engine is operating. There are a number of methods of activating hour counters:

1. **Ignition Switch.** This is the easiest and most practical method. The meter is simply connected across the ignition positive and a negative so that it operates when the engine is running.
2. **Oil Pressure Switch.** This is not common, although some installations activate through the switch so that operation is only when the engine is operating.
3. **Alternator.** In many installations the counter is activated from the alternator auxiliary terminal D+ or 61.

ACOUSTIC ALARM SYSTEMS

Acoustic alarms are generally interconnected to warning light circuits, and the buzzer is activated by a relay. Acoustic warnings are activated along with the lamp from sensor contact "W." The acoustic alarm should be activated through a relay, not through the sensor contact, which is not rated for high electrical current loads.

1. **Buzzer Test.** Using a wire bridge lead, connect a positive supply to the buzzer positive terminal and ensure that a negative one is connected. If the buzzer operates, remove the wire bridge lead. Ideally, a test function should be inserted into the circuit so that the alarm function can be verified.

2. **Operating Test.** With the alarm lights on, put a bridge wire from negative to the buzzer negative, as sometimes a "lost" negative is the problem. Connect a positive supply to the relay positive, typically numbered 86. If the relay does not operate and the buzzer is working, the relay is possibly inoperative. Verify this after removal using the same procedure. Note that sometimes a relay may sound like it is operating but the contacts may be damaged and open-circuited. If a buzzer is not operating along with lights, either a wire or connection is faulty or the operating relay is defective.

3. **Mute Function.** On many home-built engine panels, it is essential to silence the alarm, which entails placing a switch in line with the buzzer; the lamp will remain illuminated to indicate the alarm status.

4. **Time Delays.** During engine startup, a time delay is necessary to prevent alarm activation until oil pressure has reached normal operating level. Time delays typically range 15 to 30 seconds.

FUEL MONITORING

There are many factors that influence fuel consumption rates, including bottom fouling, vessel trim, propeller and engine condition, and adverse tidal flows and currents. Burning excess fuel is expensive, and optimizing consumption by varying speed to the most economical for prevailing conditions, or altering trim tab position on powerboats, is possible based on accurate fuel consumption data. Performance drops show up in fuel consumption rates, which can include propeller damage or bottom conditions. This can result in fuel consumption rate decreases of up to 20% on a vessel with a clean bottom and matched propeller, while overall savings can exceed 30% to 35%. Flow monitors use a flow sensor, some use a paddlewheel-based system, and some use an opto-electronic turbine sensor that uses an infrared light source to count turbine rotations. A processor calculates fuel flow based on the speed of the flow, and processing may use inputs from a log sensor or GPS to compute mile per gallon (mpg), liter per hour (lph), or gallon per hour (gph) rates. In systems with a return to the fuel tank, a sensor is installed in the forward and return lines, with return flow subtracted from forward flow rates.

OPEN SENSOR TEST

Remove the sensor lead marked "G" from the back of the gauge. Switch on meter supply voltage. The gauge needle should now be in the following positions:

1. Temperature gauge in the left-hand hard-over position.
2. Pressure gauge in the right-hand hard-over position.
3. Tank gauge in the right-hand hard-over position.

SENSOR GROUND TEST

This test involves the bridging of sensor input terminal "G" to negative. The sensor lead must be removed and the meter supply on. The gauge needle should be as follows:

1. Temperature gauge in the right-hand hard-over position.
2. Pressure gauge in the right-hand hard-over position.
3. Tank gauge in the left-hand hard-over position.

HOW TO TEST VDO SENSORS

Disconnect the cables and, using a multimeter, set to the resistance (Ohms/Ω) range. Place the positive (red) meter probe on the terminal marked "G" on the sensor; if a dual alarm and sensor output, the alarm output is marked "W." Place the negative (black) meter probe on the sensor thread.

Temperature Sensors

The readings should be as follows:

1. 40°C = 200Ω to 300Ω
2. 120°C = 20Ω to 40Ω

Pressure Sensors

The readings should be as follows:

1. High Pressure (engine off) = 10Ω
2. Low Pressure (engine running) = 40psi: 105Ω; 60psi: 152Ω

Fuel Tank Sensors

The readings should be as follows:

1. Tank Empty = 10Ω
2. Tank Full = 180Ω

ACKNOWLEDGMENTS

I would like to acknowledge the use of source data and information, downloaded and derived from technical data sheets; user, installation, and maintenance manuals; and the many conversations with other marine system professionals who contributed advice and information. These include the numerous marine engineers on the many ships and offshore oil rigs I served on, who taught me so much. This edition is dedicated to the memory of my marine engineer friends and colleagues who have passed due to tragic accidents while serving at sea: Chris Reed, Alan Matthews, John Trout, and Mark Paradiso.

Diesel Engines: Beta Marine, Bukh Diesel, Cummins, Caterpillar, Detroit Diesel, Dahl, FloScan, Klinger-Thermoseal, Nanni Diesel, Mercedes, Marine 16, Parker Racor, Perkins, SeaSeal, Solé Diesel, Steyr Motors, T-ISS Safety Suppliers, Trident Marine Systems, Volvo Penta, VETUS, Yanmar, Westerbeke.

Engine Electrics and Instrumentation: Olympic Instruments, VDO Marine.

Associations, Clubs, and Resources: Cruising Association (CA), Westerly Owners Association, Downunder Rally, Seven Seas Cruising Association, Island Cruising Association, Ocean Cruising Club, Noonsite, Boat Owners Association of the United States (BoatUS), Royal Yachting Association (RYA), International Marine Certification Institute (IMCI), Lloyd's Register (LR), American Boat and Yacht Council (ABYC), American Bureau of Shipping (ABS), British Marine.

BIBLIOGRAPHY

Automotive Electrics and Electronics, 4th edition. Bosch.

Engine Monitoring on Yachts. Donat, Hans. VDO Marine.

APPENDIX: MAINTENANCE CHECKLISTS

The following recommended maintenance checklists are reproduced here from chapter 7 for your convenience to photocopy for use with your vessel.

ENGINE LAY-UP MAINTENANCE TASKS

___ 1. Run the engine until warm, then drain and replace the oil. Change the filter. Drain and replace the transmission oil when warm. To avoid condensation, top up the oil.

___ 2. Top up the fuel tank, add a stabilizer to the fuel, then circulate the treated fuel and check for leaks. Drain any water within the filter/separator bowl.

___ 3. Clean the seawater strainer. Rinse the seawater system with freshwater to dilute salt, if possible.

___ 4. Check and renew the coolant additive and run the engine to circulate properly.

___ 5. Lubricate the throttle and gear change linkages using light machine oil.

___ 6. Tape over the exhaust outlet using duct tape.

ENGINE SPRING MAINTENANCE TASKS

___ 1. Change the oil before the start of the season or at the end of the lay-up period, including the oil filters. Run the engine, put in gear, and add some load against the mooring lines. Run up to normal operating temperature.

___ 2. Operate the transmission and throttle cables over the full range several times and lubricate if required.

___ 3. Check and clean the seawater strainer.

___ 4. Check the engine mounts; clean and tighten.

___ 5. Check the electrical connections: clean and tighten.

DAILY AND WEEKLY INTERVAL TASKS

___ 1. Check the lube oil level and top up as required.

___ 2. Check the fuel oil pre-filter for water and drain.

___ 3. Check coolant water level and top up as required.

___ 4. Check transmission oil level and top up as required.

___ 5. Check engine for fuel, oil, and water leaks.

___ 6. Check and top up battery electrolyte levels.

150-HOUR/1-MONTH INTERVAL TASKS

___ 1. Check the fuel filter and replace as required.

___ 2. Check the air filter.

___ 3. Check and clean the crankcase breathers.

___ 4. Perform lubricating oil analysis.

___ 5. Check water system anodes if installed.

___ 6. Check and drain water tanks.

___ 7. Check and clean seawater strainers.

___ 8. Open and close all seacocks and valves five times to exercise them.

250-HOUR/3-MONTH INTERVAL TASKS

___ 1. Change the engine oil.

___ 2. Change the oil filter.

___ 3. Check the condition of the seawater pump impeller.

___ 4. Check and change transmission oil.

500-HOUR/12-MONTH INTERVAL TASKS

___ 1. Check and clean the crankcase vent tube.

___ 2. Inspect all air intake hoses and hose connections.

___ 3. Open and clean water coolers.

___ 4. Check and re-tension rubber drive belt.

___ 5. Check and tighten alternator mounting bolts.

___ 6. Check the coolant water additives.

___ 7. Check and tighten all coupling bolts.

___ 8. Check engine cable harness for chafe.

___ 9. Check and tighten starter and alternator connections.

___ 10. Check and tighten engine negative cable connections.

___ 11. Check the condition of the coolant filler cap.

1,000-HOUR/24-MONTH INTERVAL TASKS

___ 1. Check the engine crankshaft vibration damper.

___ 2. Perform a pressure test of the cooling system.

___ 3. Check and verify engine mountings are clean and tight.

___ 4. Replace the water pump impeller.

2,000-HOUR INTERVAL TASKS

___ 1. Check and adjust the valve clearances.

___ 2. Flush the cooling system and add new coolant.

___ 3. Check all injectors and fuel pumps and overhaul.

INDEX